D1800205

Measurement of crystal growth and nucleation rates

Published by
Institution of Chemical Engineers (IChemE),
Davis Building,
165–189 Railway Terrace,
Rugby, CV21 3HQ, UK

IChemE is a registered Charity

© 2002 IChemE

ISBN 0 85295 449 2

Typeset by Techset Composition Limited, Salisbury, UK
Printed by MPG Books, Bodmin, UK

Measurement of crystal growth and nucleation rates

IChem**E**

INSTITUTION OF CHEMICAL ENGINEERS

European Federation of Chemical Engineering

Working Party on Crystallization

Measurement of crystal growth and nucleation rates

Edited by

J. Garside, A. Mersmann and J. Nyvlt

Contributors:

E. J. de Jong	Delft University of Technology, NL
A. Eble	Bayer, Leverkusen, FRG
C. Gahn	BASF, Ludwigshafen, FRG
J. Garside	UMIST, Manchester, UK
C. Heyer	E.ON, Düsseldorf, FRG
M. Kind	Universitat Karlsruhe, FRG
A. Mersmann	TU München, FRG
J. Nyvlt	Academy of Sciences, Prague, CSFR
H. Offermann	RWTH, Aachen, FRG
J. Pohlisch	Degussa AG, Frankfurt, FRG
H. Schubert	Linde, München, FRG

Second Edition

Preface to the first edition

Kinetic data of crystallization are probably the most important parameters for the design of crystallizers. In most cases, these data, measured by different authors using different methods, are spread throughout the literature, so that comparison of values published by several authors in order to check the reliability of data is extremely difficult and sometimes even impossible.

The Working Party on Crystallization (WPC) in the European Federation of Chemical Engineering (EFCE) decided for this reason to summarize the knowledge of the methods for determination of crystallization kinetic data. The measurement of crystal growth rates is substantially more developed than that of nucleation kinetics and so the decision was taken to start with crystal growth rate measurements; nucleation should be treated somewhere in the future.

The object of this WPC Recommendation is to outline the most important methods of crystal growth rate measurement, contribute to the unification of methods and equipment and to show, by using worked examples, the appropriate treatment of data and their specification.

My thanks are due to Professor Mersmann (Technical University Munich) and to Professor Garside (UMIST) who undertook the difficult task of editing the publication, to the authors of individual chapters and to all the WPC Delegates for their fruitful discussions and comments.

J. Nyvlt
(WPC chairman)
Praha, April 1989

Preface to the second edition

In the 12 years since the publication of the first edition of this book, many advances have been made in the field of crystallization. We now understand much more of the fundamentals of nucleation and growth and of the role of impurities. We can better control the habit of crystals and have made great strides in our understanding of the interaction between the fluid dynamics and kinetics in determining the output from industrial crystallizers.

The importance of crystallization as a purification and separation technique has become ever more evident and the range of industrial applications is growing rapidly. Furthermore, there are increasing numbers of researchers in universities, research institutes and industry who are involved in the many aspects of crystallization research.

All these factors have emphasized the importance of reliable and reproducible kinetic measurements. This second edition of Measurement of Crystal Growth Rates extends the scope of the first edition to include the measurement of nucleation rates. Nucleation is a field where measurement is more difficult and reproducibility harder to achieve, partly of course due to the inherent stochastic nature of the nucleation process itself. Nevertheless, great strides have been made in this area over the last decade and so we have felt it timely to include nucleation rate measurements in this edition.

We hope that this book will be of value to many people, not least those starting out on research careers who need particular guidance on available measuring techniques.

We would like to acknowledge the continuing support of the Working Party on Crystallization of the EFCE. Special thanks are due to many workers in the laboratory of Professor Mersmann at the Technical University Munich who have developed many of the techniques newly described in this edition.

John Garside
Manchester, August 2002

Contents

Introduction

1

In 1982, the German 'VDI' (Verein Deutscher Ingenieure) published 'Meßanordnungen zur Bestimmung von Kristallwachstumsgeschwindigkeiten' (methods for the determination of crystal growth rates). At the National Working Party on Crystallization, Dipl. Ing. W. Wöhlk gathered various contributions with the objective of standardization and unification of methods, units and symbols.

The Working Party on Crystallization (WPC) of the European Federation of Chemical Engineers (EFCE) forms the logical authority to author a publication to standardize and unify equipment, procedures and methods. With increasing international co-operation between researchers at universities and in industry, the need for such a reference has grown significantly.

The first edition of this book was published in 1990 and was concerned solely with the measurement of crystal growth rates. This second edition, again produced under the auspices of the WPC and now published by the Institution of Chemical Engineers in the UK, extends the scope to include measurement of nucleation rates.

The aim of this book is to recommend:

- proven methods, equipment and procedures for growth and nucleation rate measurements;
- complete information on parameters relevant to the measured data under discussion;
- symbols and units for presentation and publication;

and to promote:

- international co-operation and discussion in the field of crystallization;
- training for young researchers; and
- publication of reliable data suitable for design and operation of industrial crystallizers.

While deliberating on which symbols and units to use, we broadly adhered to a list of proven expressions that have been in use for over thirty years[1,2].

It should be emphasized that researchers who have recently applied the methods in laboratories are responsible for most of the chapters in this book. In some cases this has resulted in a very detailed description of equipment and apparatus. The description may sometimes appear tedious, but this detail is imperative in order to obtain reliable and comparable results.

Fundamentals of kinetics

2

2.1 Concentration

Since crystal size distribution depends on crystallization kinetics, that are in turn strongly influenced by supersaturation, the determination of concentrations (as well as temperatures) is very important for crystallization processes.

2.1.1 Concentration units

The number of collisions of elementary particles (atoms, molecules, ions) in a fluid phase depends on the number of these particles per unit volume:

$$\frac{\text{Number of elementary particles}}{\text{Volume of fluid phase}} = \frac{n \cdot N_A}{V} = c_i \cdot N_A \qquad (2.1.1)$$

N_A is the Avogadro number and n/V the moles n in volume V. For processes which are governed by the number of collisions (for example, primary nucleation and chemical reactions) the molar concentration of component i, c_i [kmol m^{-3}] is the quantity which governs the rates (in the above cases the nucleation and reaction rates).

It is often more convenient to use the mass concentration C_i:

$$C_i = c_i \cdot \tilde{M}_i \qquad (2.1.2)$$

In addition to these concentrations, mole and mass fractions (x_i and w_i respectively) and the mole and mass ratios (X_i and W_i respectively) of two substances are used. Definitions and conversions between the various units are listed in Table 2.1.1.

Sometimes a concentration is expressed as moles of the component i per unit mass of solvent. This expression is referred to as molality M_i of the substance i. Further alternatives are mass of solute per unit mass of solvent, or mass of solute (hydrate) per unit mass of free solvent.

Table 2.1.1 Definitions and conversions of concentration units

	Referred to mass [kg] $M_i = C_i V$		Referred to amount of substance [kmol] $n_i = M_i/\tilde{M}_i$	
	Two components	k components	Two components	k components
Total mass	$M = M_a + M_b$	$M = \sum_{j=a}^{k} M_j$	Total amount of substance $n = n_a + n_b$	$n = \sum_{j=a}^{k} n_j$
Mass fraction	$w_a = \dfrac{M_a}{M}$	$w_i = \dfrac{M_i}{M}$	Mole fraction $x = \dfrac{n_a}{n}$	$x_i = \dfrac{n_i}{n}$
Mass ratio	$W_a = \dfrac{M_a}{M_b}$	$W_i = \dfrac{M_i}{M_{Carrier}}$	Mole ratio $X_a = \dfrac{n_a}{n_b}$	$X_i = \dfrac{n_i}{n_{Carrier}}$
Conversion	mass fraction w_i $\longleftrightarrow$		mole fraction x_i	
Mass fraction from mole fraction	$w_a = \dfrac{1}{1 + (\tilde{M}_b/\tilde{M}_a)((1 - x_a)/x_a)}$	$w_i = \dfrac{x_i \tilde{M}_i}{\sum_{j=a}^{k} (x_j \tilde{M}_j)}$ Mole fraction from mass fraction	$x_a = \dfrac{1}{1 + (\tilde{M}_a/\tilde{M}_b)((1 - w_a)/w_a)}$	$x_i = \dfrac{w_i/\tilde{M}_i}{\sum_{j=a}^{k} (w_j/\tilde{M}_j)}$
Conversion	mass ratio $\longleftrightarrow$ mass fraction $W = \dfrac{w}{1-w} \qquad w = \dfrac{W}{1+W}$		mole ratio $\longleftrightarrow$ mole fraction $X = \dfrac{x}{1-x} \qquad x = \dfrac{X}{1+X}$	
Mean molar mass	$\tilde{M}_m = \dfrac{M}{n} = \dfrac{1}{w_a/\tilde{M}_a + (1 - w_a)/\tilde{M}_b}$	General: $\tilde{M}_m = \dfrac{1}{\sum_{j=a}^{k} w_j/\tilde{M}_j}$	Mean density: $\rho_m = \dfrac{1}{w_a/\rho_a + w_b/\rho_b}$	

2.1.2 Concentration measurements

Concentrations can be determined by both physical and chemical measurements. It is necessary to distinguish between the preparation of a solution with a certain concentration and determination of the concentration of a solution with an unknown composition.

In the first case, it is common to weigh quantities of solvent and solute carefully and to bring them together to produce a solution by mixing. It is important to avoid any losses by solvent evaporation.

The concentration measurement of solutions with unknown compositions is much more difficult. Physical measurements based on a solution's physical properties (density, refractive index, electrical conductivity and so on) are usually used. It is also common to determine the concentration of a sample by evaporating it to dryness. There are a few substances ($KMnO_4$ or $Na_2S_2O_3$ for example), which can be analysed more accurately by chemical methods. Polythermal methods (dissolution of the last crystal with temperature rise) can also be applied for concentration measurements.

Measurements of density

Very precise data of solution density are usually needed for concentration measurements. The most widely used technique uses electronic measurements of the natural frequency, f, of a tube filled with solution. A glass tube containing the liquid solution is stimulated to undamped vibrations. The resulting natural frequency is determined by the total mass of the tube and the liquid. The latter may be either at rest or in a flowing state provided that the tube is completely filled. The natural frequency of the oscillator is transferred to a receiver.

The system, consisting of a tube of mass M and a volume of liquid V, with density ρ, is equivalent in mass to a weight supported by a spring with the constant c. The natural frequency, f, of the system is:

$$f = \frac{1}{2\pi}\sqrt{\frac{c}{M + V\rho}} \qquad (2.1.3)$$

and the period of oscillation $T = 1/f$

Thus:

$$T = 2\pi\sqrt{\frac{M + V\rho}{c}} \qquad (2.1.4)$$

with constants A and B defined as:

$$A = \frac{4 \cdot \pi^2 \cdot V}{c} \quad \text{and} \quad B = \frac{4 \cdot \pi^2 \cdot M}{c} \qquad (2.1.5)$$

5

the density is given as:

$$\rho = \frac{1}{A} \cdot [T^2 - B] \qquad (2.1.6)$$

Since A and B are apparatus constants, they can be determined by two calibration measurements of substances with known densities (for example air and water).

The difference in the densities of two fluids (1 and 2) can be expressed by:

$$\rho_1 - \rho_2 = \frac{1}{A} \cdot [T_1^2 - T_2^2] \qquad (2.1.7)$$

The accuracy of measurements carried out with density oscillators is in the range between $10^{-1}\,\mathrm{kg\,m^{-3}}$ and $10^{-3}\,\mathrm{kg\,m^{-3}}$, depending on the measuring cell and on temperature control. Since precise thermostats provide a temperature stability of about $\pm 0.01\,\mathrm{K}$, an accuracy of $\pm 1.5 \cdot 10^{-3}\,\mathrm{kg\,m^{-3}}$ can be obtained.

As an example Figure 2.1.1 shows the saturation density, ρ_L^*, of potash alum plotted against the anhydrate mass ratio, W^*. It can be shown that the obtainable accuracy of the density measurement would lead to an accuracy of $\pm 2 \cdot 10^{-3}$ g anhydrate per kg H_2O in the mass ratio.

Example
Calibration measurements were carried out with air and water at $\vartheta_M = 40°C$ (measuring temperature).

Air: $p = 980\,\mathrm{kPa}$, $\vartheta_M = 40°C \rightarrow \rho_{air} = 1.07\,\mathrm{kg\,m^{-3}}$ $T = 1{,}224{,}171$
H_2O: $\vartheta_M = 40°C \rightarrow \rho_{H_2O} = 992.21\,\mathrm{kg\,m^{-3}}$ $T = 1{,}694{,}977$

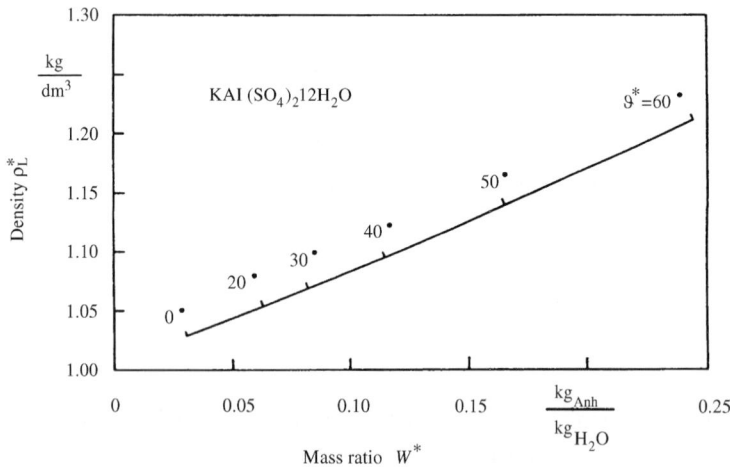

Figure 2.1.1 Density of saturated solution ρ_L^* as a function of mass ratio W^*

The apparatus constants A and B can be calculated using Equations (2.1.8) and (2.1.9) which follow from Equation (2.1.6).

$$A = \frac{T_{H_2O}^2 - T_{air}^2}{\rho_{H_2O} - \rho_{air}} = 1.386638 \cdot 10^9 \tag{2.1.8}$$

$$B = T_{air}^2 - A \cdot \rho_{air} = 1.49711 \cdot 10^{12} \tag{2.1.9}$$

The period of oscillation of a potash alum solution is measured to be $T = 1,714,930$. Using Equation (2.1.6) the density can be calculated:

$$\rho_L = \frac{1}{1.386638 \cdot 10^9} \cdot (1,714,930^2 - 1.49711 \cdot 10^{12}) = 1066.08 \, kg \, m^{-3}$$

From Figure 2.1.1 the corresponding saturation mass ratio can be calculated to be:

$W^* = 0.0775 \, kg$ anhy per kg H_2O.

Refractive index

Since the velocity of light depends on the density of the material it is travelling through, a light beam is refracted when it passes from one medium to another. The refractive index, RI, is equal to the ratio of the light velocities within the two media, or the ratio of the sines of the refractive angles:

$$RI = \frac{c_A}{c_B} = \frac{\sin \alpha_A}{\sin \alpha_B} \tag{2.1.10}$$

The value of RI is determined by refractometers. Figure 2.1.2 (see page 8) shows a typical refractometer that can measure the refractive index of a small quantity of solution at a specific temperature. Since the refractive index depends on temperature, measurements must be made at precisely fixed temperature. *In situ* measurements can be carried out by refractometers whose prisms are submerged in the solution with unknown concentration. Using differential refractometers, the difference in the refractive indices of the reference solution and the unknown solution can be determined (see Figure 2.1.3 on page 8).

The sensitivity of the refractometers represented in Figures 2.1.2 and 2.1.3 is in the range between $2 \cdot 10^{-5}$ and about $6 \cdot 10^{-6}$ for differential refractometers.

Electrical conductivity

In order to compare the conductivity of various systems, it is necessary to define some electrical quantities. The electrical resistance, R, of a body in a homogeneous electrical field is proportional to the length, l, of the body, and

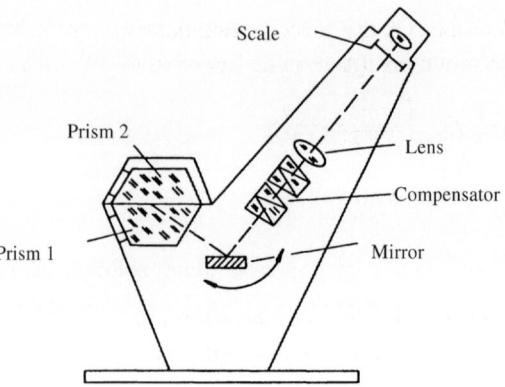

Figure 2.1.2 Refractometer

inversely proportional to the cross sectional area, A, perpendicular to the electrical field:

$$R = \rho \cdot \frac{1}{A} \ [\Omega] \tag{2.1.11}$$

The proportionality factor, ρ, is called the specific resistance:

$$\rho = R \cdot \frac{A}{1} \ [\Omega m] \tag{2.1.12}$$

while the inverse of the specific resistance is the specific conductivity, χ:

$$\chi = \frac{1}{\rho} = \frac{1}{R \cdot A} [\Omega^{-1} m^{-1}] \tag{2.1.13}$$

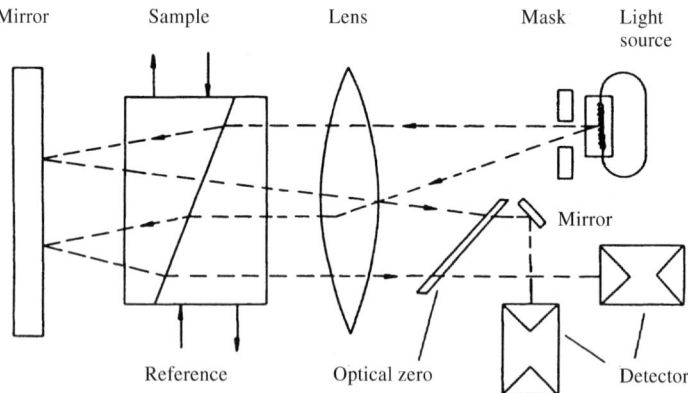

Figure 2.1.3 Differential refractometer

According to Kohlrausch, the concentration dependence of the conductivity of dilute electrolyte solutions ($A^{n+} + B^{n-}$) is often expressed as:

$$\Lambda = \Lambda_\infty - k_c\sqrt{c} \quad \left[\frac{m^2}{\text{val }\Omega}\right] \tag{2.1.14}$$

with the equivalent conductivity, Λ:

$$\Lambda = \frac{\chi \cdot n_e}{c} \tag{2.1.15}$$

n_e is the electrochemical valence of the electrolyte.

According to Figure 2.1.4, the specific conductivity of many aqueous solutions initially increases with concentration, but then passes through a maximum, because the mean free path length of ions changes with the concentration. With increasing concentration the mean free path length decreases to such a small distance that recombination of ions to molecules takes place.

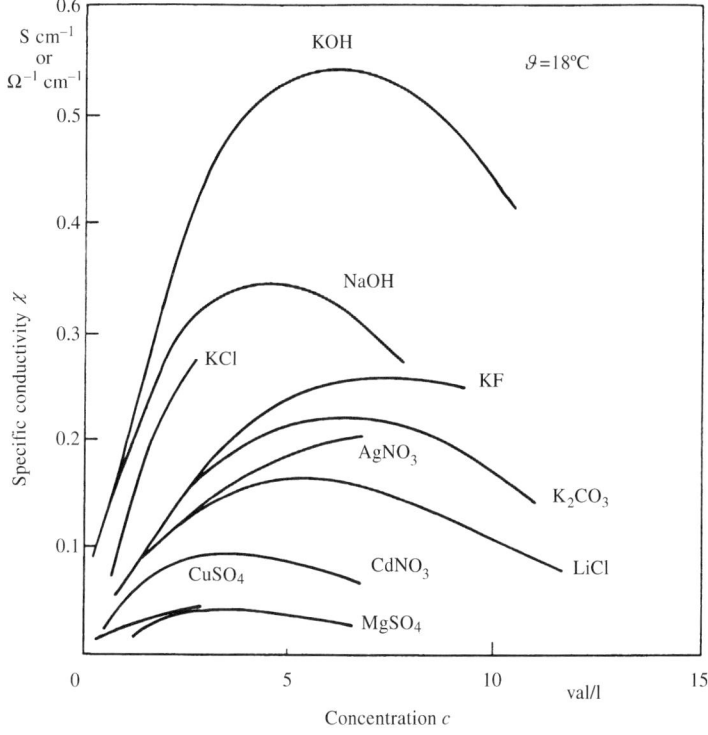

Figure 2.1.4 Electrical conductivity of different solutions

A particular disadvantage of this measuring method is that electrical conductivity is often strongly dependent on temperature, with changes of 2% to 8% per degree Celsius. In practice, the specific conductivity of a solution with an unknown concentration is compared to the known value of a calibration solution. In this way it is possible to minimize the influence of the temperature effect. Taking into account temperature stability in the range of ± 0.01 K of good thermostats, an accuracy of $\pm 0.02\%$ for $d\chi/d\theta \approx 3(1/\Omega mK)$ is attainable for KCl.

Evaporation to dryness

The easiest and most common way of measuring the concentration of a solution is to take a sample and dry it. If M_{sol} is the mass of the solution sample and M_C is the mass of the dried solid crystals after the solvent is completely evaporated, the mass ratio can be determined by the following equation:

$$W = \frac{M_C}{M_{sol} - M_C} \tag{2.1.16}$$

Measurement requires only a balance and a drying device, such as a laboratory dryer or an ultra-violet lamp, to enhance the evaporation of the solvent. It is important to take particle-free-solution samples and avoid the incursion of any dirt or dust during the drying process.

The accuracy of this method depends on sample size (the larger the sample, the smaller the expected deviation from the actual value), and on the quality of the balance (micro balances with an accuracy of 10 mg for a maximum weight of 4000 g are available).

Many inorganic crystals are rather hygroscopic. It is therefore advisable to cool the dry solids to room temperature in a desiccator in order to avoid sample weight gains from the humid air.

If a system crystallizes as a hydrate of the principal form $A \cdot z\,H_2O$ (for example, where for $MgSO_4 \cdot 7H_2O$, A is the anhydrate, $MgSO_4$ and z is the number of water molecules per molecule of anhydrate in the crystal, 7 in this case), the mass ratio can be calculated according to Equation (2.1.17):

$$W = \frac{M_{sol}}{M_C \cdot (z \cdot \tilde{M}_W)/(\tilde{M}_A + z \cdot \tilde{M}_W) + M_{W,f}} - 1 \tag{2.1.17}$$

where $\tilde{M}_A + z \cdot \tilde{M}_W = \tilde{M}_H$ is the molar mass of the hydrate.

In this equation W is the mass ratio in kg anhydrate per kg water, M_{sol} is the mass of solution being dried, M_C is the mass of crystallized hydrates (the amount of dry crystal after all the free water has evaporated), $M_{W,f}$ is the mass of free water (that evaporated during drying) and $\tilde{M}_W$, $\tilde{M}_A$ and $\tilde{M}_H$ are the relative molecular masses of water, anhydrate and hydrate, respectively.

In the case of an anhydrous material, the number, z, of bound water molecules equals zero, and Equation (2.1.17) can be simplified to:

$$W = \frac{M_{sol} - M_{W,f}}{M_{W,f}} = \frac{M_C}{M_{W,f}} = \frac{M_C}{M_{sol} - M_C} \tag{2.1.18}$$

which is the same as Equation (2.1.16).

Returning to the example of magnesium sulphate, another difficulty becomes obvious, especially if Figure 2.1.5 is considered. Some systems, like magnesium sulphate, can crystallize with different numbers of H_2O molecules per molecule of anhydrate, the number depending on the temperature at which the crystallization process takes place.

For example, 35 g of solid material crystallizing out of 100 g clear solution may be produced by different solution concentrations, depending on the

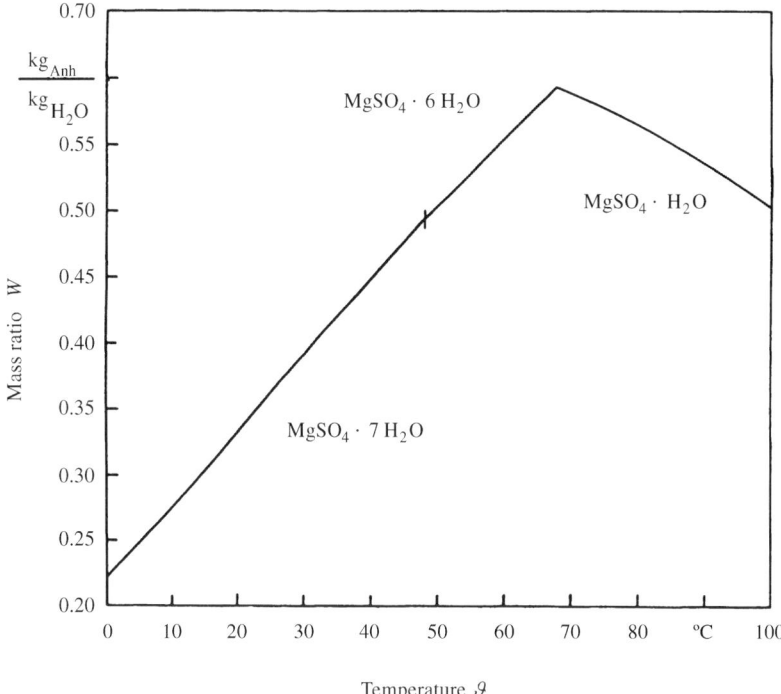

Figure 2.1.5 Mass ratio of saturated aqueous magnesium sulphate solutions as a function of temperature

11

temperature at which the solution is dried. The molecular mass of water $\tilde{M}_W$ is $18\,\mathrm{g\,mol^{-1}}$ and the molecular mass of the anhydrate of magnesium sulphate is $120\,\mathrm{g\,mol^{-1}}$.

If 35 g solids out of 100 g solution result from drying at a temperature lower than 48°C, the corresponding mass ratio can be determined as $W = 0.2059\,\mathrm{kg}$ anhydrate per kg water, as the number of water molecules, $z = 7$. For a drying temperature between 48°C and 69°C a concentration of $0.2259\,\mathrm{kg}$ anhydrate per kg water is calculated according to Equation (2.1.17), because magnesium sulphate crystallizes with $z = 6$ in this temperature range. Finally, if the drying process takes place at a temperature higher than 69°C, magnesium sulphate monohydrate is produced and the corresponding loading of the solution is $0.4375\,\mathrm{kg}$ anhydrate per kg water.

This example shows that the drying temperature has to be known if the concentration is determined by drying out a sample, especially if the system crystallizes as a hydrate. It is recommended that the fraction of water left in the hydrate after the drying process should be determined analytically to ensure correct concentration measurements.

2.2 Supersaturation

The fundamental thermodynamic driving force for a crystallization process is the difference in chemical potential between the crystallizing substance, i, in the crystal ($\mu_{i,C}$) and the solution or liquid phase ($\mu_{i,L}$), therefore:

$$\Delta\mu_i = \mu_{i,C} - \mu_{i,L} \tag{2.2.1}$$

Crystallization takes place when $\Delta\mu_i < 0$ and it is convenient to introduce the 'growth affinity', $\phi = -\Delta\mu$. With the definition of chemical potential:

$$\mu_i = \mu_i^0 + \tilde{R}\cdot T\cdot \ln a_i \tag{2.2.2}$$

where a_i is the solute activity, the fundamental dimensionless driving force for the growth process, $\sigma_{\ln(a_i)}$ can be written:

$$\sigma_{\ln(a_i)} = \frac{\phi_i}{\tilde{R}\cdot T} = v_i \cdot \ln\frac{a_i}{a_i^*} = v_i \cdot \ln\left[\frac{\gamma_i\cdot c_i}{\gamma_i^*\cdot c_i^*}\right] \tag{2.2.3}$$

where the superscript * refers to the equilibrium or saturation value (corresponding to subscript C in Equation (2.2.1)) and the subscript L (corresponding to the supersaturated solution) has been dropped for convenience; v_i is the number of ions in a molecular unit.

Equation (2.2.3) is seldom used since activities or (what amounts to the same thing) activity coefficients, γ_i, are difficult to obtain, especially for concentrated solutions. So three distinct approximations are usually made:

i) ν_i is invariably put equal to unity, which is justifiable only for non-electrolytes.

ii) It is assumed that the activity coefficient ratio, γ/γ^*, is unity. This is likely to be valid in dilute solutions and where c and c* are close together. In other words, the activity coefficients are taken to be independent of concentration over a small concentration range. Thus (omitting i for one solute)

$$\sigma_{\ln(a)} \sim \sigma_{\ln(c)} = \ln\left[\frac{c}{c^*}\right] \qquad (2.2.4)$$

For highly soluble systems, Equation (2.2.4) gives values of $\sigma_{\ln(c)}$ which depend on the concentration units employed for the solution concentrations, highlighting the need to use Equation (2.2.3).

iii) Equation (2.2.3) can be rewritten as:

$$\sigma_{\ln(a)} = \ln\frac{a}{a^*} = \ln(\sigma_a + 1) \qquad (2.2.5)$$

where

$$\sigma_a = \frac{a - a^*}{a^*}.$$

For $\sigma_a \ll 1$,

$$\sigma_{\ln(a)} = \ln(\sigma_a + 1) \sim \sigma_a \qquad (2.2.6)$$

This approximation gives an error of about 0.5% at $\sigma_a = 0.01$ and about 5% at $\sigma_a = 0.1$.

Simultaneous application of all the above approximations leads to:

$$\sigma_{\ln(a)} \sim \sigma_c = \frac{c - c^*}{c^*} \qquad (2.2.7)$$

Equation (2.2.7) is the most widely used expression for the driving force or supersaturation for crystal growth. The simplifications set out in its derivation should be noted. It is not possible to give a specific rule for the choice of concentration units to be used in this equation. Molar concentration, c (moles per unit volume), is perhaps the most satisfactory from a theoretical viewpoint, but the use of volume units causes practical difficulties because of volume changes with temperature or concentration changes during crystallization. In practice, therefore, mass or moles per unit mass of solution or solvent are usually preferred. It should be pointed out that when dealing with hydrated salts, concentrations must be expressed in terms of the hydrated salt, not the anhydrous material[75]. A more detailed discussion of the above points has been given by a number of authors[3-6].

In the correlation of crystal growth rate measurements it is common to use the concentration driving force:

$$\Delta c = c - c^*$$ (2.2.8)

as the practical driving force for the process, and this expression will be used frequently in this volume. One further frequently used expression is the supersaturation ratio:

$$S = \frac{c}{c^*}$$ (2.2.9)

As none of the quantities σ_C, Δc or S represent the fundamental driving force for crystal growth, their values change with the units employed[7] and great care should be taken to specify the units used in their evaluation. Expressions for σ_C, Δc and S are frequently expressed in terms of mass concentrations, C_i, or mass fractions, w_i, rather than molar concentrations, c_i.

2.3 Crystal growth and nucleation rates

2.3.1 Growth

Crystals grow by the advance of the individual faces present on the crystal. These faces are usually, although not always, flat. In general, each face will grow at a different rate and the relative growth rates of different faces determine crystal habit or shape. Faster growing faces tend to grow out of the crystal and so those faces making up the major part of the crystal surface are the slower growing faces.

Crystallographically similar faces (all the {1 1 1} or octahedral type faces in a potash alum crystal for example) will grow at similar rates. However, the occurrence of growth rate dispersion (see Section 2.5) means that in practice there will be variations in growth rate between such faces, and sometimes variations with time for one specific face. So it is important to define the specific growth rate that is to be measured carefully and to be clear about the reasons for measuring that growth rate. The particular growth rate that is most suitable depends on the purpose to which the measurement is to be put.

There are three main ways of expressing the growth rate of a crystal or population of crystals:

i)   Face growth rate, v_{hkl}. This is the velocity of advance of the crystallographic face, (hkl), measured perpendicular to the face and hence has the

14

same dimensions as velocity, $(m\,s^{-1})$. This is the only growth rate that can be related directly to fundamental theories of crystal growth based on mechanistic descriptions of the growth process. Examples of such theories are the Burton, Cabrera, Frank 'layer' growth theory[8] and the 'nuclei upon nuclei' or 'birth and spread' theories[9]. In order to measure v_{hkl} it is necessary to observe and make measurements on individual crystal faces.

ii) Overall mass growth rate, best expressed as the total mass flux to the crystal surface, R_G. This is a growth rate averaged over the whole crystal. Typical units of R_G are $kg\,m^{-2}\,s^{-1}$. For a crystal, mass M_C and surface area A_C, it is given by:

$$R_G = \frac{1}{A_C} \cdot \frac{dM_C}{dt} \tag{2.3.1}$$

If the face growth rates, v_{hkl}, and areas A_{hkl}, of all the faces on a crystal are known, R_G can be related to the different values of v_{hkl} by the expression:

$$R_G = \frac{\rho_C}{A_C} \cdot \sum v_{hkl} \cdot A_{hkl} \tag{2.3.2}$$

where the summation is taken over all the faces present.

The overall mass growth rate is particularly valuable for yield calculations and design purposes, particularly batch systems.

iii) Overall linear growth rate, G, is defined as the rate of change of a characteristic dimension, L, of the crystal. Thus:

$$G = \frac{dL}{dt} \tag{2.3.3}$$

and has dimensions of velocity.

The value of G clearly depends on the specific characteristic dimension used in the definition and it is vital to define this dimension (see Section 2.6.1) when using the overall linear growth rate. G and R_G can be related as follows:

$$R_G = \frac{1}{A_C} \cdot \frac{dM_C}{dt} = \frac{1}{\beta \cdot L^2} \cdot \frac{d(\alpha \cdot \rho_C \cdot L^3)}{dt} = \frac{3 \cdot \alpha \cdot \rho_C}{\beta} \cdot \frac{dL}{dt} \tag{2.3.4}$$

therefore

$$R_G = \frac{3 \cdot \alpha}{\beta} \cdot \rho_C \cdot G \tag{2.3.5}$$

15

Note that for spheres and cubes, for which $\beta/\alpha = 6$:

$$R_G = \frac{1}{2} \cdot \rho_C \cdot G \quad \text{or} \quad G = 2 \cdot \frac{R_G}{\rho_C} \tag{2.3.6}$$

The overall linear growth rate is widely used in population balance theory[10] and hence in design procedures for continuous and, to some extent, batch crystallizers.

2.3.2 Nucleation

The process of creating a new solid phase from a supersaturated mother phase is called *nucleation*. The entities that are so produced are termed *nuclei*. There are a number of very different mechanisms that produce nuclei. Nucleation in a clean, homogeneous solution that results solely from the coming together of molecular 'clusters' is known as *primary homogeneous*, or *activated* nucleation. With *primary heterogeneous* nucleation, the kinetics are catalysed by the presence of an existing surface in the form of poorly-defined 'dust'. When the surface is the crystallizing material itself, *secondary* nucleation results, induced only because of the prior presence of crystals of the material being crystallized. Secondary nucleation generally occurs at much lower supersaturations (or undercoolings) than primary homogeneous or even heterogeneous nucleation.

An example of the consequences of these different mechanisms is provided by the freezing of water. Carefully purified water can be undercooled to perhaps $-30°C$ before primary homogeneous nucleation occurs and tap water, nucleating by a heterogeneous mechanism, can be cooled to perhaps $-9°C$; a continuous crystallizer containing a retained bed of ice crystals will operate perfectly well at -2 or $-3°C$, with secondary nucleation producing the necessary number of new nuclei.

Primary nucleation is often important in the production of speciality chemicals such as dyes, pharmaceuticals, photographic chemicals, pigments and catalysts, where the solute solubilities tend to be low and so the supersaturations are high. With relatively soluble materials, secondary nucleation mechanisms dominate the production of new particles. Such a mechanism occurs when batch crystallizers are 'seeded' and is always present in a continuous crystallizer.

Secondary nucleation can occur by a number of different mechanisms. If dry crystal seeds are placed in a supersaturated solution they will often shed particles of crystalline dust that had been adhering to their surfaces. These shed particles then become new centres for growth, or 'secondary nuclei'. Crystals that grow in the form of needles or dendrites are often rather fragile and small parts of the crystal may break, again forming secondary nuclei. There is evidence that the shear forces imposed on a crystal face by the solution flowing past it can be sufficient to produce secondary nuclei from the surface, so-called *shear nucleation*.

The most important and commonly encountered mechanism of secondary nucleation is *contact nucleation*, sometimes also referred to as 'collision nucleation' or 'collision breeding'. Contacts between a growing crystal and walls of the container, a stirrer or pump impeller, or other crystals result in contact nucleation. It is now recognized that for materials of high or moderate solubility this is the most significant nucleation mechanism in crystallizers.

2.4 Growth rate expressions

Crystal growth rate expressions link crystal growth rate to supersaturation. There are many different ways of expressing both of these variables (see Sections 2.2 and 2.3) and consequently a multiplicity of growth rate expressions.

2.4.1 Empirical expressions

The simplest expression is an empirical relation between overall growth rate, expressed in terms of R_G, and the overall concentration driving force, for example ΔC. Thus:

$$R_G = k_G \cdot (C_\infty - C^*)^g = k_G \cdot \Delta C^g \qquad (2.4.1)$$

g is the overall 'order' of the growth process and k_G is the growth rate constant which will, in general, be a function of temperature, relative crystal/solution velocity and impurities within the system. Similar expressions can be written in terms of v_{hkl} and G and of other driving forces.

Equation (2.4.1) does not reflect the two separate steps involved in the crystal growth process. These are directly analogous to examples of diffusion followed by chemical reaction as studied in chemical reaction engineering. Solute must first diffuse from the bulk solution to the crystal/solution interface, after which integration of solute into the crystal lattice (or surface reaction) takes place. It can be assumed that these two processes take place in series, each being driven by a different driving force since the solution concentration at the crystal/solution interface, C_I, is intermediate between that in the bulk solution, C_∞, and the equilibrium value C^*. The value of C_I depends on the relative magnitude of the resistances offered by the diffusion and integration steps.

The solute concentration profiles in the vicinity of the crystal surface are illustrated in Figure 2.4.1.

The two-step process can best be described by the equations:

diffusion: $\qquad R_G = k_d \cdot \dfrac{(C_\infty - C_I)}{1 - w_\infty}$ $\qquad\qquad$ (2.4.2)

integration: $\qquad R_G = k_r \cdot (C_I - C^*)^r$ $\qquad\qquad$ (2.4.3)

k_d is the conventional mass transfer coefficient (dimension LT^{-1}) and can be determined directly from dissolution rate measurement (see Section 3.3.2) or from mass transfer correlations. The term $1 - w_\infty$, where w_∞ is the bulk solution concentration expressed as a mass fraction, arises from the bulk flow term in the diffusion equations[11].

Elimination of the unknown interface concentration, C_I, between the above two equations gives:

$$R_G = k_r \cdot \left[(C_\infty - C^*) - R_G \cdot \frac{1 - w_\infty}{k_d} \right]^r \qquad\qquad (2.4.4)$$

This expression allows the surface integration kinetics to be determined from overall growth rate measurement so long as the mass transfer coefficient is known.

If the integration process is 'first order' ($r = 1$), then Equation (2.4.4) can be rearranged to give:

$$R_G = k_G \cdot (C_\infty - C^*) \qquad\qquad (2.4.5)$$

where

$$\frac{1}{k_G} = \frac{(1 - w_\infty)}{k_d} + \frac{1}{k_r} \qquad\qquad (2.4.6)$$

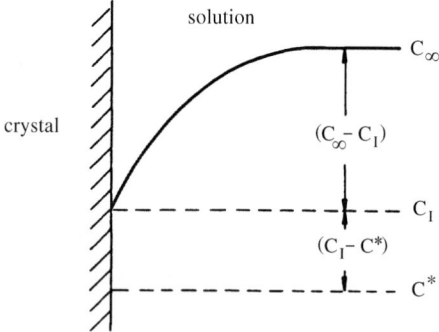

Figure 2.4.1 Concentration gradients adjacent to a crystal surface

Equation (2.4.5) is thus a special case of Equation (2.4.1). If $r \neq 1$ or $r \neq 2$, the general growth rate equation (Equation 2.4.4) cannot be rearranged in the form of Equation (2.4.1), emphasizing the empirical nature of this latter equation. For $r = 2$, the following equation is obtained[12,13].

$$R_G = k_d \cdot \Delta C + \frac{k_d^2}{2 \cdot k_r} - \sqrt{\frac{k_d^4}{4 \cdot k_r^2} + \frac{k_d^3 \cdot \Delta C}{k_r}} \qquad (2.4.7)$$

2.4.2 Diffusion and integration control

The development in Section 2.4.1 demonstrates how the 'resistance in series' model leads to the possibility that either of the two steps may control the overall growth rate. Thus if $k_d \gg k_r$, the diffusion step offers far less resistance than the integration process and $k_G \approx k_r$. In this case, the concentration driving force over the boundary layer, $C_\infty - C_I$, is much smaller than the force driving the integration process, $C_I - C^*$, so the interface concentration will approach the bulk concentration and the concentration profile will be similar to that shown in Figure 2.4.2. For this case:

$$R_G \approx k_r \cdot (C_\infty - C^*)^r \qquad (2.4.8)$$

This process is said to be 'integration controlled'. The opposite case, 'diffusion control' occurs if $k_r \gg k_d$.

The concept of effectiveness factors, well established in reaction engineering, is useful in characterizing the extent to which the diffusional resistance

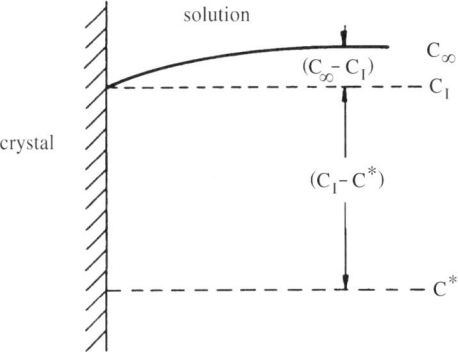

Figure 2.4.2 Concentration gradients adjacent to a crystal surface for the case of integration controlled growth

influences the growth rate. A crystal growth rate effectiveness factor, defined by the expression:

$$\eta_C = \frac{\text{growth rate at the interface conditions}}{\substack{\text{growth rate that would be obtained if the interface} \\ \text{were exposed to the bulk conditions}}} \qquad (2.4.9)$$

is given by[14,15]:

$$\eta_C = (1 - \eta_C \cdot Da)^r \qquad (2.4.10)$$

where the Damköhler number for crystal growth is defined by:

$$Da = k_r \cdot (C_\infty - C^*)^{r-1} \cdot \frac{(1 - w_\infty)}{k_d} \qquad (2.4.11)$$

and represents the ratio of pseudo-first order rate coefficient at the bulk conditions to the mass transfer coefficient.

Equation (2.4.10) is plotted in Figure 2.4.3. When Da is large, growth is diffusion controlled and $\eta_C \to Da^{-1}$. Conversely when Da is small $\eta_C \to 1$ and growth is controlled by the integration step.

The effect of changes in driving force and in mass transfer coefficient are illustrated in Figure 2.4.4 (see page 21). Growth rates have been calculated from Equation (2.4.4) for a second order integration step ($r = 2$). Typical

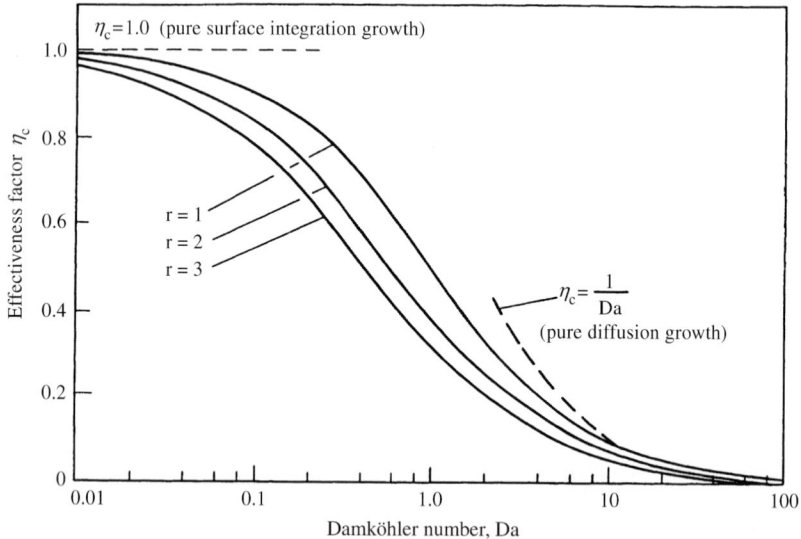

Figure 2.4.3 The effectiveness factor for crystal growth

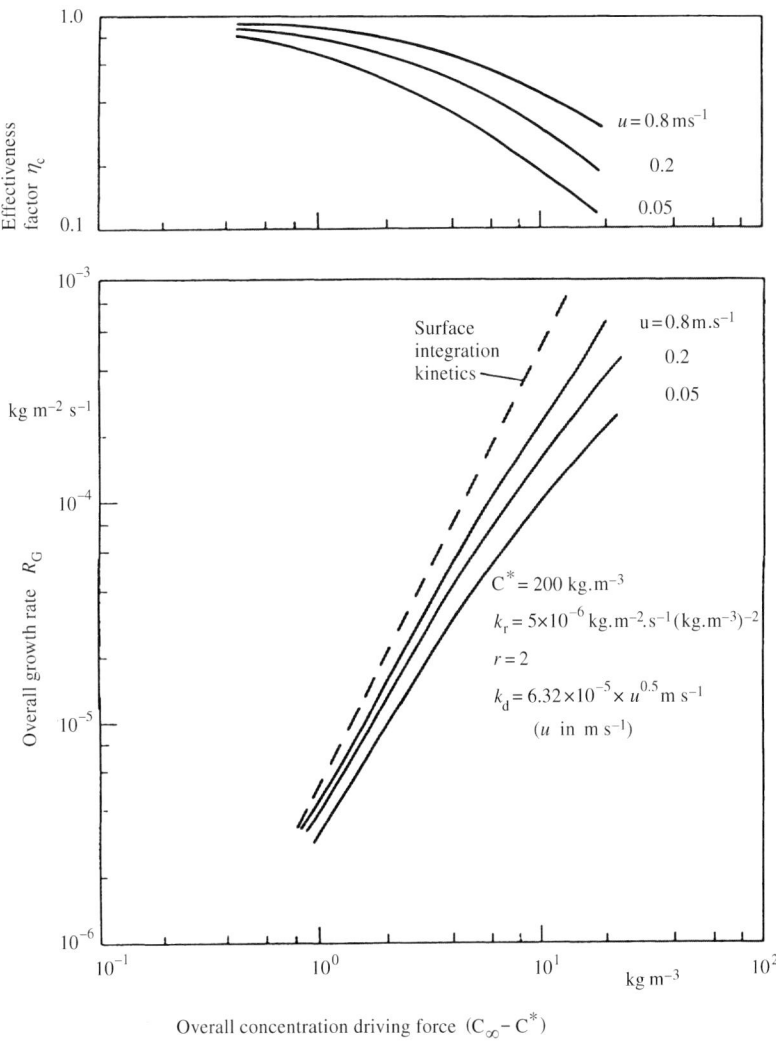

Figure 2.4.4 Influence of concentration driving force and mass transfer coefficient on overall growth rate

values have been assumed for the other parameters and are indicated in the figure. The mass transfer coefficient is assumed to vary with the relative crystal/solution velocity as $k_d \sim u^{0.5}$. Since the integration process increases more rapidly with driving force than does the diffusion step, growth rates tend to greater diffusion control at higher driving forces. This is clearly shown by the

21

decrease in the effectiveness factor. Conversely, as the relative crystal/solution velocity increases, growth becomes more integration controlled and the effectiveness factor tends to unity. This is clearly seen when growth rate is plotted as a function of relative velocity (Figure 2.4.5). Note that the curvature in the log-log plots of growth rate against driving force highlights the somewhat unsatisfactory nature of Equation (2.4.1) in expressing the overall growth process.

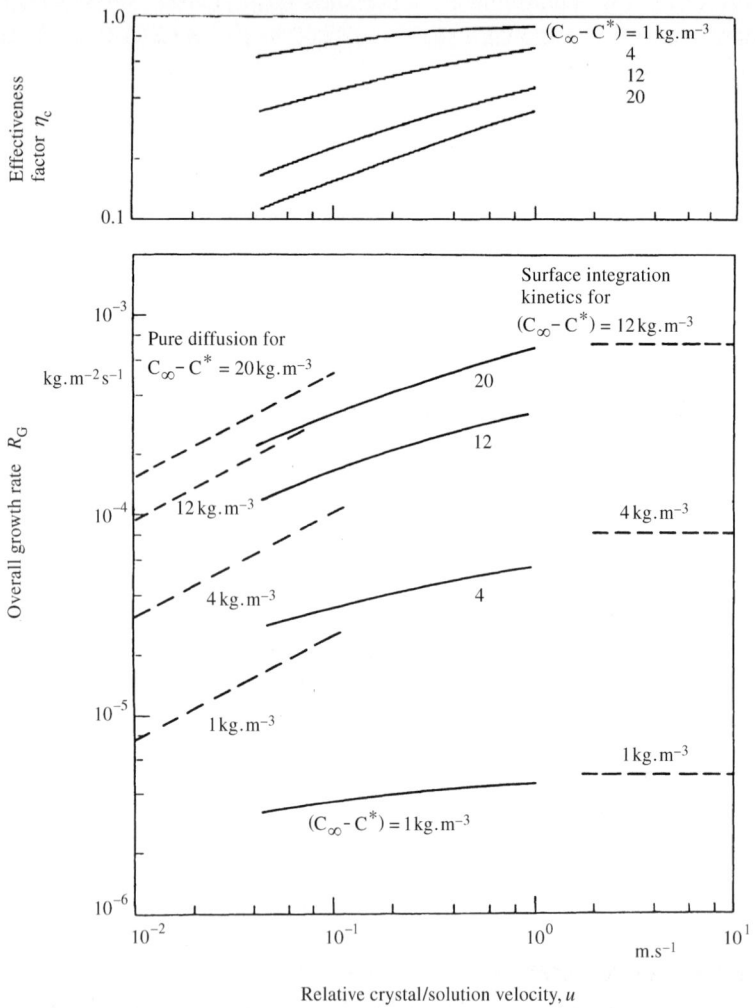

Figure 2.4.5 Influence of relative crystal/solution velocity on overall growth rate

2.4.3 Fundamental growth rate expressions

The above Equations are generally satisfactory for design purposes. If more detailed information on growth rate mechanisms is required, the empirical expression for the integration kinetics (Equation 2.4.3) must be replaced by one of the theoretical equations that describe specific integration mechanisms. Two such expressions have been widely used. The first arises from the Burton, Cabrera, Frank (BCF) theory[8] which assumes that growth takes place via 2-dimensional diffusion of growth units over the crystal surface to a step in the crystal lattice. The presence of many 'kink' sites in such a step provides easy integration of growth units into the lattice. In its simplest form, the BCF theory predicts that the face growth rate, v, is related to the supersaturation at the surface, σ_1, by:

$$v = \frac{C \cdot \sigma_1^2}{\sigma_c} \cdot \tanh \frac{\sigma_c}{\sigma_1} \qquad (2.4.12)$$

where C and σ_c are parameters related to the crystal and the surface diffusion process.

For $\sigma_1 \ll \sigma_c$ (at low supersaturations) this equation reduces to:

$$v = \frac{C}{\sigma_c} \cdot \sigma_1^2 \qquad (2.4.13)$$

while when $\sigma_1 \gg \sigma_c$ (high supersaturations):

$$v = C \cdot \sigma_1 \qquad (2.4.14)$$

These two limiting cases can, therefore, be described by the empirical relation Equation (2.4.3).

The second widely used growth expression is that describing the 'birth and spread' or 'nuclei upon nuclei' model. If stable two-dimension nuclei form on the crystal surface, growth can proceed either by the addition of many similar nuclei, or through the attachment of growth units to the edge of the nuclei via a surface diffusion flux. The equation:

$$v = A_1 \cdot \sigma^{5/6} \cdot \exp\left(-\frac{A_2}{\sigma}\right) \qquad (2.4.15)$$

seems to give a satisfactory representation of such a process[9]. The form of Equations (2.4.12) and (2.4.15) is shown in Figure 2.4.6 (see page 24).

2.4.4 Effect of temperature

Both the diffusion and integration steps are temperature dependent and this dependency can be characterized by corresponding activation energies.

23

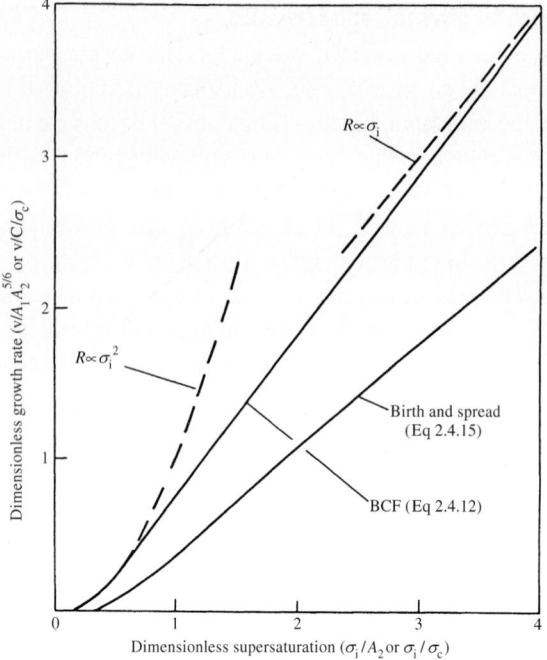

Figure 2.4.6 Form of the relation between growth rate and supersaturation predicted by the BCF and Birth and Spread theoretical growth rate expressions

Thus the mass transfer coefficient, k_d, (Equation 2.4.2) and the integration rate coefficient, k_r, (Equation 2.4.2) can be written:

$$k_d = k_{do} \cdot \exp\left(\frac{-E_d}{\tilde{R}T}\right) \qquad (2.4.16)$$

and

$$k_r = k_{ro} \cdot \exp\left(\frac{-E_r}{\tilde{R}T}\right) \qquad (2.4.17)$$

where E_d is the activation energy for the diffusion step and E_r that for the integration process.

E_d is usually in the range 8–20 kJ per mol while values of E_r are invariably higher, typically 40–60 kJ per mol. As a result, the rate of the integration process increases much more rapidly with temperature than does the diffusion step and so crystal growth rates are more likely to be diffusion controlled at high temperatures and integration controlled at low temperatures. This is illustrated in Figure 2.4.7 (see page 25) where growth kinetics have again

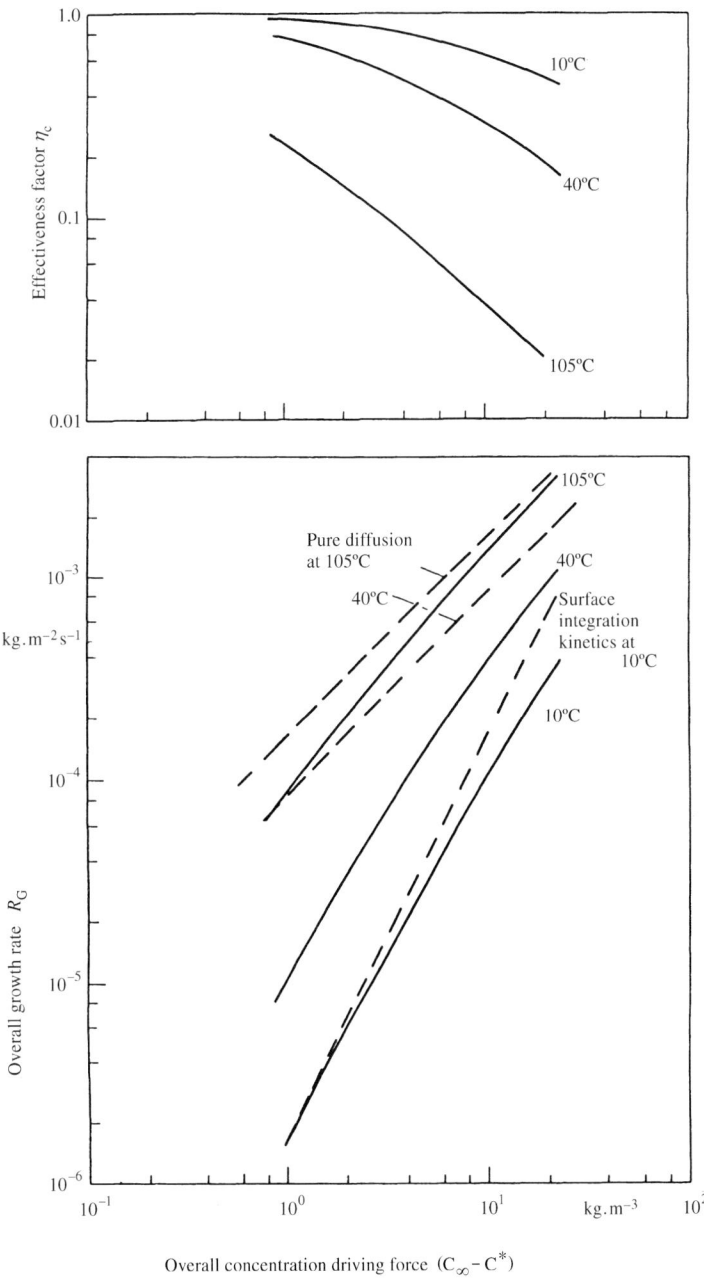

Figure 2.4.7 Effect of temperature on growth rate

25

been calculated from Equation (2.4.4) using a second order integration process ($r = 2$) and other typical values for the various parameters appearing in the equation. Changes in the controlling step are clearly seen from the variation in the effectiveness factor.

2.5 Size dependent growth and growth rate dispersion

Crystal growth rates of a given substance can vary in a number of different ways, causing complications in the interpretation of growth rate kinetics. Two main phenomena can be distinguished as size dependent growth and growth rate dispersion. These two effects, their possible cause and consequences will be discussed.

2.5.1 Size dependent growth

Size dependent growth arises when the growth rate of crystals depends on their size. The term will be restricted to the case where *all* crystals of a given size grow at the same rate, this rate being different from the growth rates of crystals of other sizes. Three possible causes for such effects have been suggested.

i) *Mass transfer:* The mass transfer coefficient to a particle suspended in a fluid depends on the particle size L as:

$$k_d = c_1 \cdot L^{-1} + c_2 \cdot L^{-0.17} \qquad (2.5.1)$$

where the first term on the right hand side of the equation arises from molecular diffusion and the second from forced convection[16]. The molecular diffusion term dominates for small particles, say less than about 10 μm in water-like solutions, while forced convection is the major term at sizes greater than about 100 μm. Mass transfer limitations with freely-suspended crystals, therefore, lead to growth rates that decrease with increasing size. Such effects are rarely seen.

ii) *Gibbs–Thomson (or Ostwald–Freundlich) effect:* The solubility, c_L, of a small crystal is related to its diameter, L, by the Gibbs–Thomson equation:

$$\frac{c_L}{c^*} = \exp \frac{4 \cdot \tilde{M} \cdot \gamma_{CL}}{\tilde{R} \cdot T \cdot \rho_C \cdot L} \qquad (2.5.2)$$

The equilibrium solubility thus increases with decreasing crystal size; crystals growing in a given solution will experience a lower supersaturation and hence a lower growth rate with decreasing size.

The crystal size at which the Gibbs–Thomson effect becomes significant depends on the surface energy γ_{CL}. Although this quantity is not known with any great accuracy, estimates indicate that the variation of solubility with size is insignificant until crystals are smaller than a few micrometers[17]. In general the Gibbs–Thomson effect is thus not a contributing factor to the size-dependent growth of crystals larger than say 10 μm.

iii) *Size-dependent surface integration kinetics:* There is some evidence that the dislocation structure within a crystal may change as a function of its size; in particular the dislocation density may increase with size. For example, mechanical stress and the incorporation of impurity atoms into the lattice may both be important factors in initiating dislocations. Further, the energy of crystal collisions and the probability of collisions in suspension systems will be greater with larger sizes. These more energetic collisions may increase both the amount and severity of damage to the crystal surface[16]. Increasing dislocation density can increase the surface integration kinetics if this is controlled by the presence of screw dislocations as in the Burton–Cabrera–Frank Theory (Section 2.4)[17].

2.5.2 Growth rate dispersion

There is now substantial experimental evidence that, when exposed to constant external conditions of supersaturation, temperature and hydrodynamics, different crystals of a given material and a given size grow at different rates. This is known as growth rate dispersion. It is useful to distinguish between 'permanent dispersion' where the dispersion in growth rates is maintained as crystals grow, a given crystal always having the same growth rate, and 'stochastic dispersion' in which individual crystals experience random fluctuations in growth rate[18]. The difference between these two models and size dependent growth are illustrated in Figure 2.5.1 (see page 28).

As with size-dependent growth, a number of factors could produce growth rate dispersion, but the most important are almost certainly those related to the surface integration mechanism. Specific examples would be variations between crystals, and variations with time for a given crystal, of

- dislocation structure;
- the presence of varying degrees of strain and deformation;
- the adsorption and/or incorporation of impurity atoms, molecules or ions; and
- the presence of different crystallographic faces.

Besides these effects agglomeration of crystals can be very important. It can be difficult to distinguish between growth and agglomeration.

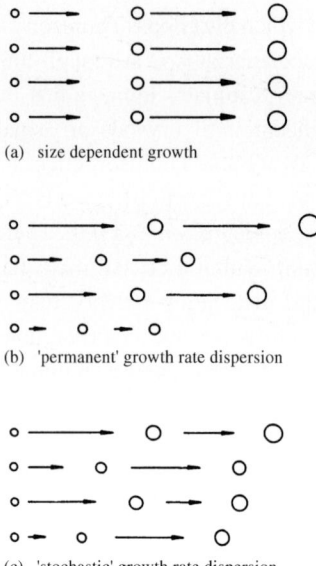

(a) size dependent growth

(b) 'permanent' growth rate dispersion

(c) 'stochastic' growth rate dispersion

Figure 2.5.1 Diagramatic representation of size-dependent growth and growth rate dispersion[18]

2.6 Shape factors

2.6.1 Definition

With the exception of isometric crystals, different faces of a crystal exhibit different growth rates resulting in non-isometric shaped crystals. It is thus necessary to characterize the shape of crystals using shape factors, particularly when overall growth rates are considered. Such shape factors serve for calculation of the crystal mass (or volume):

$$M_C = \alpha \cdot \rho_C \cdot L^3 \tag{2.6.1}$$

and crystal surface area:

$$A_C = \beta \cdot L^2 \tag{2.6.2}$$

from a characteristic dimension L. This characteristic dimension may be defined in many different ways depending on the method of size (or size distribution) measurement. For its proper definition it is necessary to start with a crystal in a standard orientation.

The basic orientation of a crystal is that of its greatest mechanical stability (Figure 2.6.1a on page 29). The length, L_a, is then the distance between two

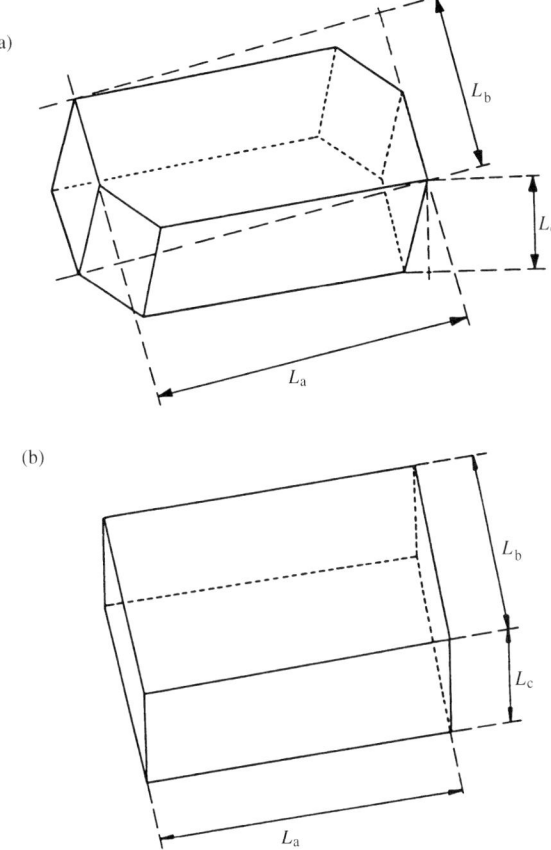

Figure 2.6.1 (a) Crystal in state of greatest mechanical stability and (b) approximation to this crystal shape

parallel planes perpendicular to the base, contacting the crystal in such a way as to give maximum length. The width, L_b, is the distance between two planes which are perpendicular both to the base and to the planes defining the length, L_a and contacting the crystal on opposite corners, edges or planes. The thickness, L_c, is the distance between the base and another parallel plane which contacts the crystal from above.

From the sieve analysis of nearly isometric particles, the value of the sieve aperture, L_s, (the minimum sieve aperture through which the crystal can just pass) may be used for crystal size characterization. For crystals that are reasonably isometric, L_s is approximately equal to the second largest dimension of the crystal:

$$L_s \approx L_b \tag{2.6.3}$$

If the size of crystals is determined by optical analysis (for example, micro-scopy, photomicroscopy or image analysis) the measured value can be the projected area diameter, L_p, i.e. the diameter of a circle having the same area as a projection of the crystal perpendicular to the plane of greatest stability. Thus:

$$L_p = \left(\frac{4 \cdot L_a \cdot L_b}{\pi}\right)^{1/2} \tag{2.6.4}$$

According to Allen[19] this projected area diameter is related to the sieve diameter by the empirical relation:

$$L_p = 1.40 \cdot L_s \tag{2.6.5}$$

For needle-shaped crystals it is convenient to take the length L_a as the characteristic dimension.

From a sedimentation analysis there results the Stokes' diameter, L_{St}. This is the diameter of a sphere having identical sedimentation velocity, u, and density as the crystal in the same liquid:

$$L_{St} = \left(\frac{18 \cdot u \cdot \eta}{g \cdot \Delta\rho}\right)^{1/2} \tag{2.6.6}$$

It is thus clear that the value of the shape factors is affected by the definition of the characteristic crystal dimension. In addition, individual particles in a polycrystalline sample have their own individual shape factors; even if an average shape factor for individual narrow size fractions is determined, these values can differ for various crystal sizes.

There are many definitions of shape factors in the literature[7,19-21], but these will not be detailed here. For the purpose of crystal growth rate data evaluation the following definitions will be used:

volume shape factor $\qquad \alpha_a = \dfrac{V_C}{L_a^3}$ $\qquad\qquad\qquad$ (2.6.7a)

or $\qquad\qquad\qquad\qquad \alpha_b = \dfrac{V_C}{L_b^3}$ $\qquad\qquad\qquad$ (2.6.7b)

surface shape factor $\qquad \beta_a = \dfrac{A_C}{L_a^2}$ $\qquad\qquad\qquad$ (2.6.8a)

or $\qquad\qquad\qquad\qquad \beta_b = \dfrac{A_C}{L_b^2}$ $\qquad\qquad\qquad$ (2.6.8b)

overall shape factor $\qquad F_a = \dfrac{\beta_a}{\alpha_a}$ $\qquad\qquad\qquad$ (2.6.9a)

or $\qquad\qquad\qquad\qquad F_b = \dfrac{\beta_b}{\alpha_b}$ $\qquad\qquad\qquad$ (2.6.9b)

Another useful parameter characterizing crystal shape is sphericity, ψ. This is defined as the ratio of the surface area of a sphere having the same volume as the crystal to the actual crystal surface area. It can be shown that:

$$\psi = \frac{(6 \cdot a/\pi)^{2/3}}{\beta/\pi} \tag{2.6.10}$$

For isometric particles ψ is close to 1, while for needle-shaped crystals or platelets its value is much lower. Evaluation of ψ is useful for checking the values of α and β since $0 \leq \psi \leq 1$.

2.6.2 Determination of shape factors

(i) The crystal shape may be approximated by the shape of a similar simple geometric body. Thus Figure 2.6.1b (see page 29) represents an approximation to the crystal shape illustrated in Figure 2.6.1a. Calculation of shape factors is then performed using Equations (2.6.1) and (2.6.2) or by making use of Table 2.6.1 (see Figure 2.6.2).

(ii) The volume shape factor can be determined by weighing N_{tot} crystals selected from a narrow sieve fraction; α_b is then given by:

$$\alpha_b = \frac{M_C}{N_{tot} \cdot \rho_C \cdot L_b^3} \tag{2.6.11}$$

where L_b is given by Equation (2.6.3) and M_C is the total mass of the N_T crystals. The surface shape factor can then be calculated according to Matz[21]:

$$\beta_b = 2 + 4\alpha_b \tag{2.6.12}$$

Table 2.6.1 Examples of shape factors (see Fig. 2.6.2)

Geometric shape	α_a	β_a	F_a	α_b	β_b	F_b	ψ
a) Sphere	0.524	3.142	6.00	0.524	3.142	6.00	1.00
b) Tetrahedron	0.118	1.732	14.68	0.182	2.309	12.7	0.68
c) Octahedron	0.471	3.464	7.35	0.471	3.464	7.35	0.85
d) Hexagonal prism	0.867	5.384	6.21	2.60	11.20	4.31	0.82
e) Cube	1.000	6.000	6.00	1.000	6.000	6.00	0.81
f) Needle $5 \times 1 \times 1$	0.040	0.88	22	5	22	4.40	0.64
g) Needle $10 \times 1 \times 1$	0.010	0.42	42	10	42	4.20	0.53
h) Plate $10 \times 10 \times 1$	0.100	2.40	24	0.10	2.4	24	0.43

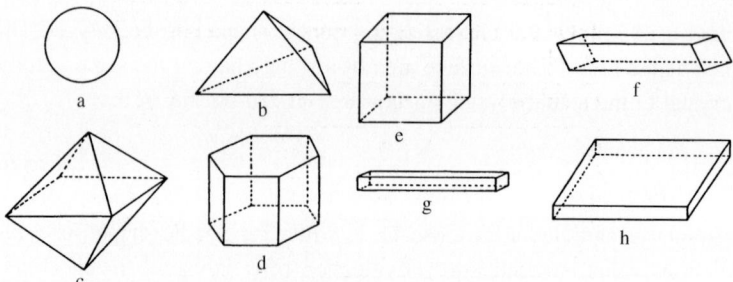

Figure 2.6.2 Idealized crystal shapes (see Table 2.6.1)

Alternatively, if the size is determined from microscopic measurements, length L_a can be estimated and then:

$$\alpha_a = \frac{M_C}{N_{tot} \cdot \rho_C \cdot L_a^3} \qquad (2.6.13)$$

and

$$\beta_a = 6 \cdot \alpha_a^{2/3} \qquad (2.6.14)$$

(iii) Matz[21] determined the average volume shape factor for a polycrystalline sample as follows: For small samples (for example 0.1 g) of individual sieve fractions, L_i, the corresponding number of crystals, N_{Ti}, was determined. A plot of log $N_{tot,i}$ versus log L_i (see Figure 2.6.3) should then give a straight line of slope -3 and intercept on the log N_{tot} axis equal to log $(M_C/\alpha_b \cdot \rho_C)$. The shape factor α_b can thus easily be calculated. If the plot cannot be approximated by a straight line, the intercept of a tangent at an arbitrary point would determine the shape factor for that particular crystal size. The surface area shape factor β_b can subsequently be estimated using Equation (2.6.12).

The consistency of the values of α and β should be checked using the sphericity (Equation (2.6.10)) and values in Table 2.6.1.

The following is an example of this method:

The numbers of crystals corresponding to samples each of 0.1 g are given in Table 2.6.2.

The slope is close to -3. From the intercept $= \log(M_C/\alpha_b\rho_C)$

$$\frac{M_C}{\alpha_b \cdot \rho_C} = 1.523 \cdot 10^{-7}$$

32

Table 2.6.2 Numbers of crystals corresponding to samples of 0.1 g

$L \cdot 10^3$ (m)	N_{tot}^*	log N_{tot}^*	log L
0.2	18855	4.28	−3.70
0.4	2357	3.37	−3.40
0.5	1206	3.08	−3.30
0.7	440	2.64	−3.15
1.0	151	2.18	−3.00

*These numbers, each corresponding to a mass of 0.1 g, were determined by weighing between 50 and 100 crystals.

With a crystal density $\rho_C = 2286 \, \text{kg m}^{-3}$

$$\alpha_b = \frac{1 . 10^{-4}}{(1.523 \cdot 10^{-7} \cdot 2286)} = 0.287$$

and from Equation (2.6.12)

$$\beta_b = 2 + (4 \cdot 0.287) = 3.15$$

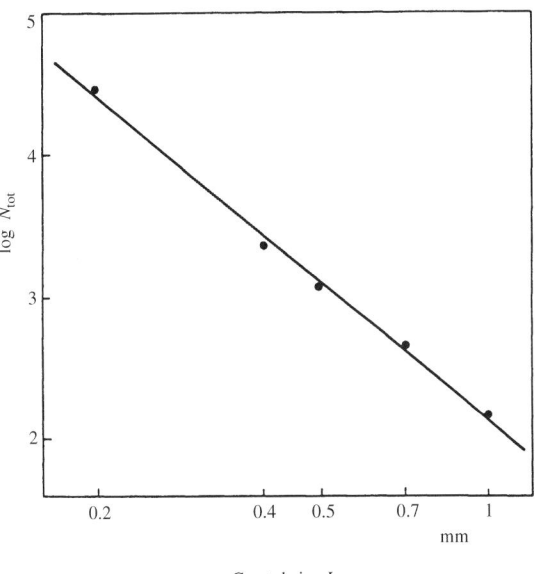

Crystal size L

Figure 2.6.3 Shows a plot of log N_{tot} against log L and using the method of least squares

The sphericity calculated from Equation (2.6.10) is then $\psi = 0.67$. Comparing these three values with similar values in Table 2.6.1 we can conclude that they seem to be reasonable. Microscopic measurement gave the values of $\alpha_b = 0.288$ and $\beta_b = 3.01$.

Fundamental equations demonstrating the use of shape factors in crystal growth rate measurements can be summarized as follows[22]:

The crystal surface area:

$$A_C = \frac{\beta_b}{\alpha_b^{2/3}} V_C^{2/3} = \frac{F_b \cdot V_C}{L_b} \qquad (2.6.15)$$

The linear growth rate in the a direction:

$$G_a = \frac{V_{Ct}^{1/3} - V_{C\alpha}^{1/3}}{2 \cdot \alpha_a^{1/3} \cdot t} \qquad (2.6.16)$$

if α_a is independent of crystal size. For a size dependent shape factor:

$$G_a = \frac{1}{2 \cdot t} \cdot \left[\left(\frac{V_{Ct}}{\alpha_{a,t}} \right)^{1/3} - \left(\frac{V_{C\alpha}}{\alpha_{a,\alpha}} \right)^{1/3} \right] \qquad (2.6.17)$$

The subscripts α and t denote the values for seeds (start) and for crystals growing during the time period, t, respectively.

The overall linear growth rate:

$$G = \frac{6 \cdot G_a}{F_a} = \frac{6 \cdot G_b}{F_b} \qquad (2.6.18)$$

2.7 Size distributions

2.7.1 Introduction

An industrial crystallizer is expected to produce solids having not only a certain composition but also a certain size distribution. The crystal size distribution (CSD) is important in determining easy and efficient separation of the mother liquid from the crystals, low residual moisture, low costs for separation and drying, avoiding the risk of dust formation during drying, and in influencing bulk properties such as density and angle of response.

It is clear that the CSD is one aspect of a crystallizer operation that demands attention, and so the development of methods for measuring the CSD are vitally important. Therefore, some attention will be paid to the principles of size measurements together with discussion of the different ways of defining and presenting size distributions.

2.7.2 Definition of size distributions

It is often difficult to indicate the size of a crystal as crystals may exhibit a large variety of shapes, for example cubic, spherical, needle-like and platelets. It is possible, however, to define a characteristic crystal length or diameter which can then be related to crystal surface areas or volume using shape factors (see Section 2.6).

The 'average' size is an important property of an assembly of crystals. The purpose of an average is to represent a group of individual values in a simple and concise manner. It is important, therefore, that the average should be representative of the group. There are different ways of defining the average size, all of which should be a measure of the central tendency, unaffected by the relatively few particles in the tail of the distribution. This will be illustrated in a number of graphs which together survey those commonly used in particle technology.

In the horizontal axis of Figure 2.7.1 the size classes are plotted. These are usually determined by the measuring device (for example aperture of a set of sieves). On the vertical axis is the relative percentage number frequency, q_0, of crystals; this is the fraction of the 'counted' crystals belonging to a certain size class (expressed as a percentage) divided by the width of the size class:

$$q_0(L_i) = \frac{N_i}{\sum_i N_i} \cdot \frac{1}{\Delta L_i} \tag{2.7.1}$$

The use of the relative percentage has the advantage that the values are independent of the number counted. The most commonly occurring size, the *mode*, corresponds to the maximum of the relative frequency curve. The

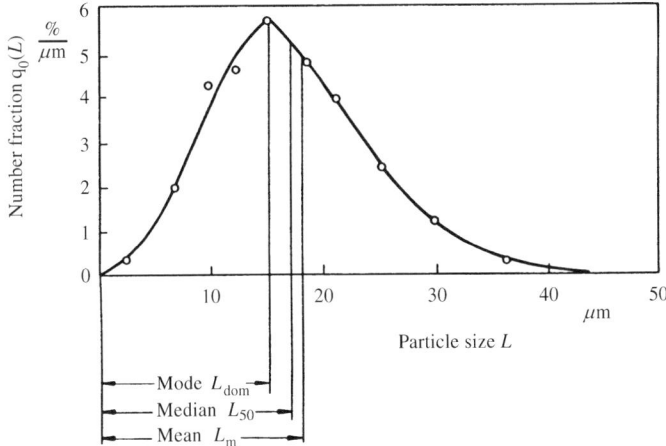

Figure 2.7.1 The relative percentage frequency curve

35

median size is such that it divides the area under the curve into equal parts, i.e. it is the 50% size on the cumulative frequency curve. The vertical line at the mean size passes through the centre of gravity of the area under the curve. The mode and the median may be determined graphically; the *mean*, however, has to be calculated.

The size distribution from Figure 2.7.1 can also be presented in a cumulative way, (Figure 2.7.2) using the equation:

$$Q_0 = \int_{L_{min}}^{L} q_0 \, dL = \sum_{L_{min}}^{L} q_0 \cdot \Delta L_i = \frac{\sum_{L_{min}}^{L} N_i}{\sum_{L_{min}}^{L_{max}} N_i} \tag{2.7.2}$$

Figure 2.7.2 shows the cumulative undersize and represents the percentage (or fraction) of the measured crystals being smaller than a certain value (for example, 50% of the measured crystals are smaller than about $17 \, \mu m$). In contrast to the cumulative undersize curve, the cumulative oversize distribution $1 - Q_0$ is sometimes plotted, this giving the percentage of crystals larger than a certain value.

Crystal size is not always the key value and it is also possible to plot the percentage of crystals having a surface or a volume and mass smaller than a certain value. For example by comparison with Equation (2.7.1):

Relative percentage surface frequency:

$$q_2(L_i) = \frac{A_i}{\sum_i A_i} \cdot \frac{1}{\Delta L_i} = \frac{L_i^2 \cdot N_i}{\sum_i L_i^2 \cdot N_i} \cdot \frac{1}{\Delta L_i} \tag{2.7.3}$$

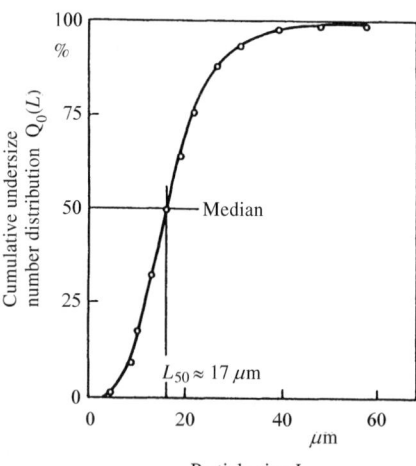

Figure 2.7.2 The cumulative percentage frequency curve

Relative percentage volume frequency:

$$q_3(L_i) = \frac{V_i}{\sum_i V_i} \cdot \frac{1}{\Delta L_i} = \frac{L_i^3 \cdot N_i}{\sum_i L_i^3 \cdot N_i} \cdot \frac{1}{\Delta L_i} \qquad (2.7.4)$$

If the density, ρ_C, is constant for all crystals, $q_3(L_i)$ is equal to the relative percentage mass frequency.

By integration of these relative percentage frequencies, corresponding cumulative graphs can be calculated. The choice of crystal number, surface, volume or mass depends on the measuring technique. A change in the vertical axis can only be made if the different shape factors are known.

The various mean crystal sizes L_m are given by:

$$\text{Mean size on length basis} = \frac{\sum L \cdot \Delta N}{\sum \Delta N} \qquad (2.7.5)$$

$$\text{Mean size on surface basis} = \sqrt{\frac{\sum \beta \cdot L^2 \cdot \Delta N}{\sum \Delta N}} \qquad (2.7.6)$$

$$\text{Mean size on volume basis} = \sqrt[3]{\frac{\sum \alpha \cdot L^3 \cdot \Delta N}{\sum \Delta N}} \qquad (2.7.7)$$

in which β is the surface shape factor, α the volume shape factor and ΔN the number counted in a certain fraction. N is the number of crystals per unit volume (m^3) of slurry. Taking slurry volume as the reference has the advantage that it immediately follows from the sample volume. If a unit of clear liquid volume is chosen, this has to be calculated from the slurry volume with the aid of the volume percentage of crystals.

The mode (or dominant size, L_{dom}) may be evaluated by differentiating the corresponding distribution function with respect to L and equating the result to zero. For example the mass distribution function can be written as:

$$W(L) = \alpha \cdot \rho_C \cdot L^3 \cdot \frac{dN}{dL} = \alpha \cdot \rho_C \cdot L^3 \cdot n \qquad (2.7.8)$$

where W(L) is the mass of the crystals in the size class between L and $L + dL$ and $n = dN/dL$ the population density. This is equal to the slope of the line in the cumulative graph on a number basis. Differentiating and equating to zero leads to:

$$\frac{dW(L)}{dL} = 3 \cdot \alpha \cdot \rho_C \cdot L^2 \cdot n + \alpha \cdot \rho_C \cdot L^3 \cdot \frac{dn}{dL} = 0 \qquad (2.7.9)$$

L_{dom} can then be calculated if $N = f(L)$ (the size distribution) is known.

In addition to the different average sizes, the width of the distribution around the mean size can be characterized by the *coefficient of variation* C.V., equal to the standard deviation divided by the mean size. For a given mass distribution the C.V. can be determined from:

$$\text{C.V.} = \left[\left[\frac{\int_0^\infty (L - L_{50})^2 \cdot W(L)\,dL}{\int_0^\infty W(L)\,dL}\right]^{1/2}\right] \cdot \left[\frac{\int_0^\infty W(L)\,dL}{\int_0^\infty L \cdot W(L)\,dL}\right] \quad (2.7.10)$$

$$\underbrace{\qquad\qquad\qquad\qquad}_{\text{standard deviation}} \qquad \underbrace{\qquad\qquad\qquad}_{(\text{mean size})^{-1}}$$

The standard deviation and coefficient of variation can only be calculated if N or n are known as a function of L.

When the cumulative size percent curve is a straight line on arithmetic-probability co-ordinates, thus approximating a Gaussian distribution, Equation (2.7.10) can be simplified to Equation (2.7.11), where $L_{16\%}$, $L_{50\%}$ and $L_{84\%}$ may be taken from a cumulative undersize distribution curve:

$$\text{C.V.} = \frac{L_{16\%} - L_{84\%}}{2 \cdot L_{50\%}} \quad (2.7.11)$$

Canning and Randolph[23] have pointed out that very few crystallizing systems exhibit a true Gaussian distribution. They suggested that the crystal size data should not be averaged by drawing a straight line, as the experimental results are then forced into a Gaussian distribution, but proposed that $L_{16\%}$, $L_{50\%}$ and $L_{84\%}$ should be taken from a curve drawn through the original data points.

Figure 2.7.3 (see page 39) shows three different cumulative undersize mass distributions analysed by sieving. The median and mean crystal size as well as the coefficient of variation can be compared.

2.7.3 Graphical presentation of size distribution

It is common practice to plot size distributions in such a way that a straight line results, with all the advantages that follow from such a reduction. This can be done if the distribution fits a certain law.

In the field of powder technology the following laws may be applicable.

The arithmetic normal distribution

This distribution occurs when the measured value of some property of a system is determined by a large number of small effects, each of which may or may not operate. If a large number of measurements of the value is made, and the results are plotted as a frequency distribution, the well-known bell-shaped curve results. In a cumulative graph the required straight line is obtained, from which the mean size can be deduced.

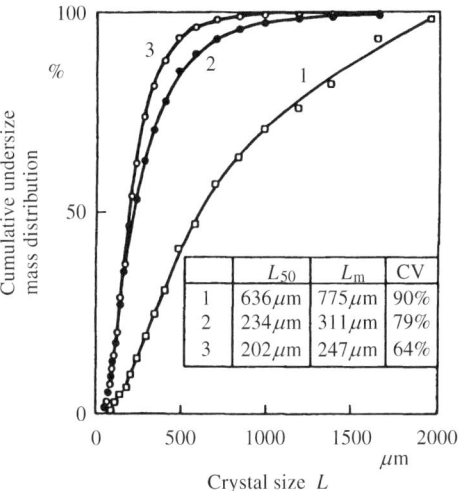

	L_{50}	L_{m}	CV
1	$636\,\mu m$	$775\,\mu m$	90%
2	$234\,\mu m$	$311\,\mu m$	79%
3	$202\,\mu m$	$247\,\mu m$	64%

Crystal size L

Figure 2.7.3 Cumulative mass distributions

The log normal distribution

According to the normal distribution, there are differences of equal amounts in excess or deficit from a mean value which are equally likely. With the log normal distribution there are ratios of equal amounts which are equally likely. In order to obtain a symmetrical curve of the same shape as the normal curve, it is necessary to plot the relative frequency against the log size. These graphs are less accurate than those of the normal law.

The Rosin Rammler Sperling Bennet (RRSB) distribution

This distribution function has been found to apply to many granular materials. The function is based on the empirical relation:

$$M(L) = \exp\left[-\left(\frac{L}{L_s}\right)^{\gamma}\right] \qquad (2.7.12a)$$

where $M(L)$ ($= 1 - Q_3(L)$ for $\rho_C = $ const.) is the weight percentage for example retained on a sieve of aperture L, L_s is the size of crystals corresponding to $100\%/e = 36.8\%$ of the product oversize fraction and γ is the so-called classification exponent, whose value often lies within 1.5–6 for crystalline material. A graph of log-log $M(L)$ plotted against log L gives a straight line. The treatment is useful for monitoring operations leading to highly skewed distributions, but should be used with caution since the device of taking loglogs always reduces scatter and is not recommended for this reason.

With respect to sieve analysis the Γ distribution function is very advantageous:

$$\Gamma(x) = \int_0^\infty \exp(-L)L^{x-1}\, dL \qquad (2.7.12b)$$

In general all three distribution functions mentioned here suffer a serious deficiency; they do not take into account the laws which govern the formation of a crystal size distribution such as nucleation (at small sizes) and their subsequent growth. To find a relation between the population density n (or number N) and the crystal size L, a number balance has to be formulated for a certain size range ΔL.

Crystals grow into this size interval ΔL; crystals are also withdrawn from the interval via the product outlet and crystals grow out of that interval. In the steady state there will be no accumulation of the number of crystals in the interval. For the simple case in which crystals are born at zero size and grow according to a very simple growth model (among others no agglomeration, no attrition, etc.), then Equation (2.7.13)

$$n = n_0 \cdot \exp\left(-\frac{L}{G \cdot \tau}\right) \qquad (2.7.13)$$

can be derived. n_0 is the population density at zero size of the crystals, G is the growth rate and τ the average crystal residence time in a spatial evenly mixed crystallizer with no product classification. To obtain a straight line for the distribution curve ln n must be plotted versus L. (More detailed information is given in Chapter 3.)

To illustrate the advantage of a population density plot in contrast to RRSB graphs, Figure 2.7.4 (see page 41) shows population densities n (left side) for KCl at 20°C and different suspension densities $m_T = 30\ \mathrm{kg\,m^{-3}}$ (below) and $m_T = 200\ \mathrm{kg\,m^{-3}}$ (above) plotted against the crystal size L. The straight line 1 corresponds to a residence time $\tau = 30$ min and a growth rate $G = 6.7 \cdot 10^{-8}\ \mathrm{m\,s^{-1}}$, line 2 to $\tau = 90$ min and $G = 3.3 \cdot 10^{-8}\ \mathrm{m\,s^{-1}}$. These distributions are transferred to an RRSB plot, seen on the right side, which results in curves. Straight lines with equal gradients but different suspension densities m_T in a ln n–L plot, result in one curve in RRSB-nets, and a loss of information.

2.7.4 Measurement of crystal size

The complexity of sizing has long been recognized; particle shape, density, area, irregularity, agglomeration and friability are all factors which must be considered. Many crystal size distributions generated in industrial practice cover a range from critical nucleus size to a few millimetres. It is clear that

Crystal size distribution of KCl

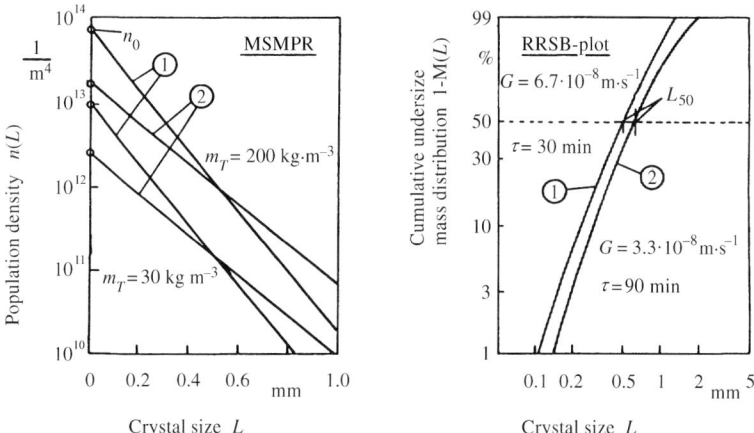

Figure 2.7.4 Comparison of a population density plot (left side) and a cumulative undersize distribution in a RRSB-net (right side)

the lower size limit of such distributions is outside the range which can be detected by existing sizing methods.

Measurement of crystal size is based on mechanical separation techniques or on the physical properties of the two phase flow of liquids and solids such as electrical and optical conductivity, Fraunhofer diffraction and image analysis.

a) Mechanical techniques

Sieving: Probably the most simple and most widely used method of determining particle size distribution. Until the introduction of micromesh sieves, the principal drawback to sieve analysis with woven sieves was that they could not be used for particles smaller than 45 μm because of the high cost of producing sieves with uniform small apertures. Micromesh sieves are made by electroforming nickel into mesh having precise openings ranging from 5 to 1200 μm. Sieving is tedious and time-consuming in particle sample preparation and sample drying.

An example of sieve analysis is shown in Section 3.5. About 10 to 15 sieves with openings between 150 μm and 2000 μm and more can be used to analyse a sample of about 1 l. A reasonable increase in successive sieve opening can be in the ratios $\sqrt[4]{2}$. The sample volume depends on the volume of the crystallizer and should not be much more than a tenth of the crystallizer volume.

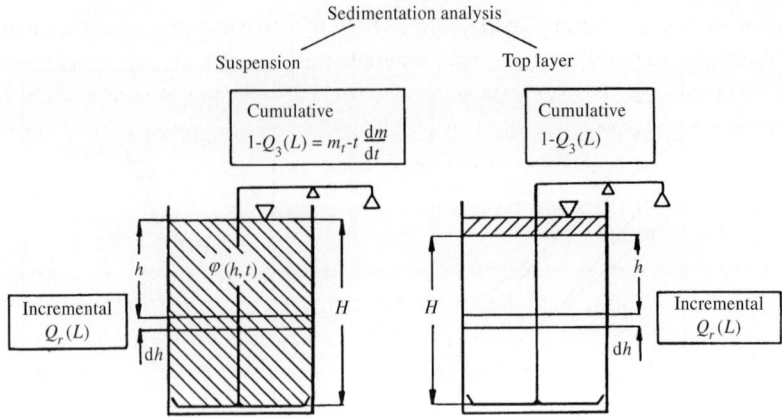

Figure 2.7.5 Sedimentation analysis

For shaking the sieves, a shaker with both horizontal and vertical movement is preferable. The time and intensity of shaking are a compromise between the demands of good separation and little mechanical stress to the crystals.

Sedimentation analysis (see Figure 2.7.5): Particles settling in a fluid reach a constant settling velocity after a short time. This effect, caused by gravitational or centrifugal forces, is used for sedimentation analysis, where the size of a particle is calculated from the measured settling velocity. The settling velocity can be measured incrementally or cumulatively starting with a suspension or with a clear fluid with a layer of particles at the top.

Choice of the fluid in which particles are dispersed for sedimentation analysis depends on a number of factors:

- chemical and physical reaction of the fluid with the dispersed particles has to be avoided, especially crystallization and dissolution.
- agglomeration must not take place.
- the fluid density must be smaller than the solid density.
- the viscosity of the fluid must be such that the analysis can be done within a reasonable time, so that large particles do not fall through the fluid faster than they can be measured. Analysis should not take too long.

Laminar flow is assumed around the particles and so the concentration must be small (smaller than $2 \cdot 10^{-3}$ vol. %) and temperature gradients have to be small to avoid convective flows.

Sedimentation balances or photo sedimenters are the most usual devices at present. A sedimentation balance weighs cumulatively the mass of solids that fall on the balance during a certain time. The undersize mass distribution can be obtained by graphical or numerical differentiation. The photo sedimenter works incrementally and measures the change in intensity of a light beam through the suspension. According to the Lambert–Beer law, the size distribution can be calculated if the coefficient of light extinction is known.

b) Physical techniques

Sensing zone methods (for example Coulter counting method) (see Figure 2.7.6): These instruments determine the total number of cumulative volume oversize distribution of particles suspended in a suitable electrolyte, by causing them to pass through a small aperture on either side of which is immersed an electrode. The changes in aperture resistance as particles pass through the sensing zone generate voltage pulses, the amplitudes of which are proportional to the particle volumes. The pulses are converted to the number of crystals present in different size classes (size channels). This method is typically used for particles between 1 and 100 μm. Each measuring cell aperture has a dynamic range of about 15:1.

The channel size, L_{ch}, represents the diameter of a sphere having the same volume as the particle being sized, i.e.

$$L_{ch} = \left(\frac{6 \cdot V}{\pi}\right)^{1/3}$$

(2.7.14)

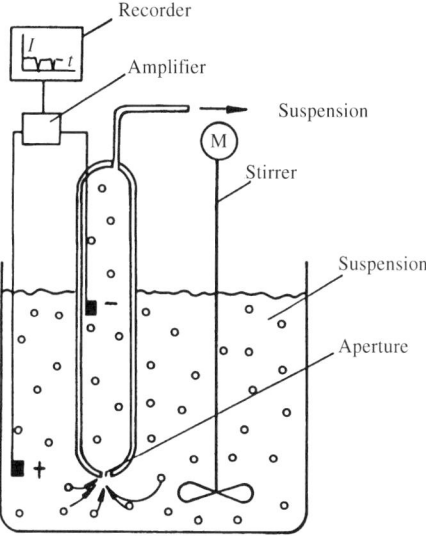

Figure 2.7.6 Principle of the sensing zone technique

Sieve analysis data are expressed in terms of a characteristic sieve aperture size. Comparison between the two sizes can be illustrated by the potash alum system. Potash alum crystals are near perfect octahedra, only (1 1 1) faces being present, and in this case the length of an edge defines the sieve size L_s. The volume shape factor corresponding to L_s is $\sqrt{2}/3$ and so:

$$\frac{\sqrt{2}}{3} \cdot L_s^3 = \frac{\pi}{6} \cdot L_{ch}^3 \qquad (2.7.15)$$

therefore

$$L_s = 1.036 \cdot L_{ch} \qquad (2.7.16)$$

To make a comparison between the two sizing methods, narrow size fractions containing potash alum crystals retained between two adjacent sieves were dispersed in saturated solution and sized by a Coulter Counter. Typical distributions for two such sieve cuts are presented in Figure 2.7.7. This shows good agreement between the average sizes obtained by the two size analysis techniques.

Optical methods: Optical instruments can cover a fairly wide size range and require very little time to perform a size analysis. Within specified particle concentration limits optical methods can be used for on-line measurement. However, size data based on optical properties must be referred to a calibration

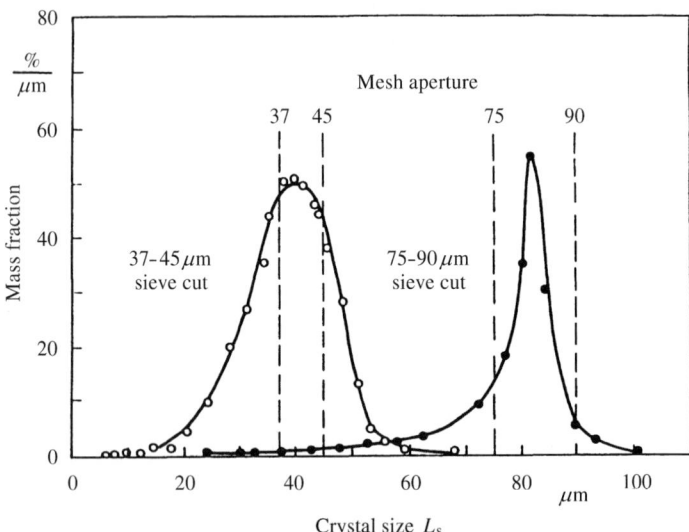

Figure 2.7.7 The relative mass percentage frequency distributions determined for two sieve cuts of potash alum crystals by the Coulter Counter analysis

base and this can involve ambiguities in cases where calibration material is not of the same kind as the particles to be sized. Furthermore, optical methods may necessitate sample dilution due to concentration restrictions and they are subject to bias if air bubbles or immiscible liquids are present. Commercially available optical instruments fall into two general classes:

i) An assemblage of particles is scanned to reveal information on overall concentration. With more sophisticated instruments of this type, spatial filtering of scattered light permits determination of the mean diameter and variance of particle size distributions.

ii) An optical system views a volume so small that individual particles in a flow are observed and each one provides an indication of size. Both concentration and size data can be derived, since the instrument reports the number of particles over several size ranges measured in a given volume of liquid.

Fraunhofer diffraction: A Fraunhofer diffraction analyser (see Figure 2.7.8) uses forward scattering of laser light. Intensity measurement is made of the Fraunhofer diffraction pattern produced by passing helium neon laser light through a suspension of crystals. While the diffraction pattern through a monodisperse suspension would take the form of coaxial Airy rings, the patterns through a polydisperse suspension is monotonic. Each particle scatters light in the forward direction with an intensity and at an angle which depends on its size. A smaller particle will defract the laser beam at a larger angle than will a larger particle. Further the overall energy level at any point varies with concentration. The diffracted beam is sensed by a photodetector which observes the light level at a series of radial points out from the beam axis to define the energy level distribution in the diffraction patterns.

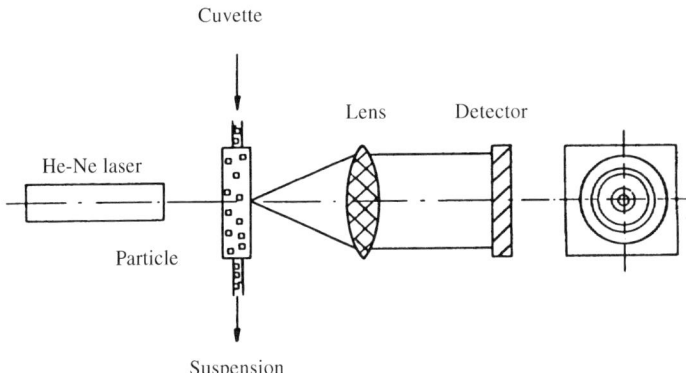

Figure 2.7.8 Fraunhofer diffraction analyser

A pre-programmed matrix calculation then permits definition of the particle size distribution. Calculations are performed digitally within the instrument's microprocessor. With the single particle counting optical devices, any transparent liquid can be used. The particle size range that can be measured with such a device is typically between 2 and 1000 μm and measurements are typically made up to almost 1% by volume. Sample liquid flow rates up to one litre per minute can be used without difficulty. Data output is normally in the format of printout giving the particle size distribution in terms of volume percentage and population per channel. Particle concentration and various mean sizes of the distribution are also available.

Light pulses: Another method of particle size analysis is counting and evaluating pulses, when light is reflected or shadowed by particles. A suspension of particles is pumped through a measuring cell. Each particle passing through a volume which is defined optically, reflects light to a measuring device or interrupts the light beam. The intensity and length of these pulses give information about particle size and concentration. The set-up of the measuring device may be in line with the light beam to evaluate the shadow or may be at an angle to the light beam (e.g. perpendicular as shown in Figure 2.7.9) if the scattered or reflected light is measured. Optical photon correlation spectroscopy may be used for particle sizes down to 3 nm. By using lenses with various focal lengths, the range of the instrument may cover a wide size distribution. The dynamic range for the shadowing device can be more than 1:100. Unfortunately, particle concentration is limited depending on the actual size

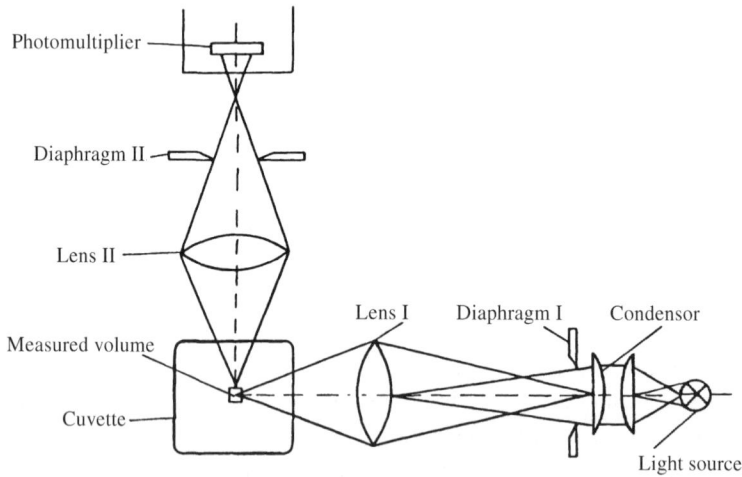

Figure 2.7.9 Measuring reflected light pulses

distribution. This is due to the fact that this method estimates the Brownian motion of particles, which depends in the case of dilute solutions only on solution viscosity, temperature and particle size but as concentration increases interparticle and hydrodynamic forces have to be taken into account, making the data analysis very hard or even impossible. For spherical particles the mathematical process for this method of analysis works straightforwardly and satisfactorily. However, crystalline materials often have different shapes and rough surfaces, which can significantly influence the distribution of scattered light and therefore complicate the analysis.

These instruments analyse particles down to a size of about 1 μm with a dynamic range of about 25:1. The dynamic range for the shadowing device, which measures the time of transition, can be more than 100:1. Particle concentration is limited to about 10^5 ml^{-1} depending on the actual size distribution.

Particle shape imposes severe limitations on this and on many other analysing methods. The mathematical processes work satisfactorily for spherical particles. Crystalline materials, however, exhibit structured surfaces which make for many difficulties. Therefore it is advisable to make reference measurements with another system and compare the results.

There are also instruments that observe a telecentric white-light beam and detect reduction in the light level caused by particle passage. With a series of sensors, particles from 1 μm to a maximum of 9 mm can be detected. Each sensor has a dynamic range reported to be 60:1, based on the smallest detectable particle (as referred to an equivalent optical diameter for a latex sphere) and the minimum dimension of the internal passageway. The smallest sensor handles particles 1 to 60 μm, in concentrations up to some 30,000 ml^{-1} at a flow rate if 4–6 ml min^{-1}. The largest handles particles up to 9 mm in concentrations up to 0.2 ml^{-1} at a flow rate of 100 l min^{-1}.

Imaging particle counters: Many manufacturers produce image analysis systems which are basically video-camera devices focused on a microscope stage or a photograph. Signals from the video image are transmitted to a computer for data reduction. A wide range of information can be developed by the computer, including length, area and size data, shape factors and statistical analysis. These instruments are very expensive and can only handle samples with a low solid concentration.

The two most important aspects of image analysis are first to prepare the sample so that the particles are distributed as single particles without any agglomeration, and second to measure a very large number of particles in order to achieve statistical significance.

In crystallizing systems the particles are generally suspended in a solution and so a representative sample must be filtrated for image analysis. Special handling is necessary to avoid difficulties in the microscope. For example membrane filters which can be made transparent, can be used. The filter paper must be clear of (super-) saturated solution in order to avoid crystallization at drying.

2.7.5 Sample preparation

Very few methods to determine particle size distributions can be performed in-line. In-line measurements, of course, do not require special preparation as the sensor for sending and receiving a signal can be placed inside the crystallizer (some light-scattering methods, which can give indications for mean sizes, or the coefficient of variation, and ultrasonic measurements, which are not yet experienced completely). Care has to be taken to find a location, where particle size distribution and sample concentration are representative of the whole crystallizer volume.

For off- and on-line measurements where a sample is removed from the crystallizer it is important to place the withdrawal tube properly and choose a velocity in that suction tube to match the local velocity of the suspension in order to get a representative sample (see Section 3.5).

On-line analysis does not usually require separation of the solids, but is often very sensitive to the crystal concentration. The suspension is pumped out of the crystallizer through a cuvette where the analysis takes place. It is often the optical density of the suspension that makes continuous measurements impossible so that Fraunhofer diffraction measurements and laser scanning methods are restricted to low suspension densities (maximum $20 \, kg \, m^{-3}$ to $50 \, kg \, m^{-3}$, depending on the particle size distribution), much smaller than are found in industrial processes. Dilution with saturated solution or clear solution from the crystallizer can solve this problem[24,25].

Separation of solids is mostly necessary for off-line analysis. This can be done for example by wet sieving as shown in Section 3.5. However, the crystals obtained by this method continue to grow as the drying solution evaporates from the crystal surface. The surface becomes imperfect, which can make further analysis with other instruments impossible. Another way to separate the solids from the solution is to filter the suspension and displace the solution still adhering on the crystal surface with an inert, viscous and immiscible fluid, for example by vacuum- or pressure-filtration. The filter cake can then be washed with another inert fluid of low boiling temperature, dried and further analysis of the crystals is then possible. This is a possible procedure for dry-sieving, image analysis and most of the other scanning, scattering and sensing methods.

2.8 Nucleation rate expressions

By analogy with growth rate expressions, nucleation rate expressions link the nucleation rate to the supersaturation and other process parameters. The nucleation rate, B, is generally defined as the total number of particles N_{tot} generated in a certain volume ΔV of constant supersaturation Δc and in a certain time Δt during which Δc remains constant:

2.8.1 Primary nucleation

The rate equation for primary nucleation is conventionally derived by assuming that clusters of individual molecules form in solution. A quasi-equilibrium develops between monomers and clusters, with a corresponding distribution of free energy as a function of cluster size. This free energy is made up of contributions from the bulk and the surface molecular free energies.

It can be shown that a maximum in this free energy occurs at a critical cluster size, corresponding to the size at which further growth of the cluster leads to a decrease in free energy. For sizes less than this, a decrease in free energy can only be achieved by dissolution. Clusters of this critical size are thus called *critical nuclei* and the chance of forming nuclei of this size will depend on the height of the free energy barrier. As the supersaturation increases the height of the barrier and the value of the critical size both decrease. With increasing supersaturation the barrier eventually becomes small enough for nucleation to become spontaneous.

The *rate* of nucleation is defined as the rate at which clusters grow through this critical size and so become crystals. This may be written in a simplified form that highlights the effect on nucleation rate of supersaturation, temperature and interfacial tension:

$$B = A_B \exp\left(\frac{-B_B \gamma_{CL}^3}{T^3 \sigma^2}\right) \qquad (2.8.1)$$

γ_{CL} is here the surface energy, T the temperature and σ the supersaturation. Figure 2.8.1 (see page 50) shows the form of this $B(\sigma)$ relationship. At low supersaturations the interfacial tension term dominates and there is insufficient free energy available via the supersaturation to create new surface, hence $B \approx 0$. At some critical value of the supersaturation, σ_{crit} the nucleation rate increases catastrophically, eventually reaching a maximum value. This increase explains the experimentally known transition from a metastable zone in which, although supersaturation exists, nucleation rates are very low, to the labile zone in which nucleation is spontaneous.

Full details of the derivations referred to above can be found in many texts[26,27].

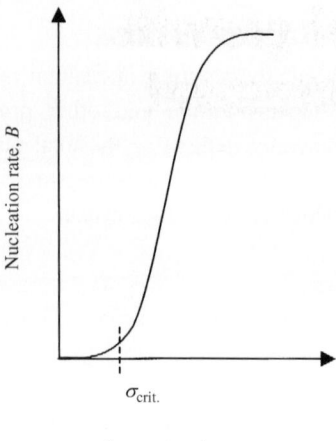

Figure 2.8.1 Form of the primary nucleation rate equation

2.8.2 Secondary nucleation

Many experiments have shown that the nucleation rate B in crystallizers depends not only on the supersaturation but also on the concentration of crystals in suspension, m_T, and on some measure of the hydrodynamic interactions between the crystals and the solution, for example stirrer speed, N, or average power input to the solution, ε. The effect of these variables is usually expressed empirically as a power law function, for example:

$$B = k_b m_T^j N^k \Delta c^b \tag{2.8.2}$$

In this form of the equation, N (a stirrer, or pump impeller, rotation rate) is taken as the measure of the fluid mechanics interactions and of the power input. Alternative formulations use the specific power imput ε to represent the influence of the fluid mechanics.

Typical values of b lie between about 1 and 2.5. These are much lower than the values that would be found if the primary nucleation rate equation (Equation 2.8.1) were fitted to a power law function of Δc. The influence of crystal concentration points directly to the importance of crystal collisions. Most values of j are close to unity, suggesting the dominance of collisions between crystals and the vessel walls or, more probably, between crystals and the stirrer rather than between two crystals. Specific values of k predicted by various semi-empirical models of contact nucleation in crystallizers are in the range 2 to 4, as are the majority of experimentally determined values.

Fuller details of the use of these equations and the mechanisms underlying the kinetics are given in subsequent chapters.

Methods of growth rate measurement

3

3.1 Introduction

The various growth rate measurements described in this Chapter can be distinguished from each other and categorized in a number of different ways. First, measurements can be made either on single crystals, or on a population consisting of large numbers of crystals. The former are particularly valuable for fundamental studies of growth mechanisms and habit modification, while the latter are usually employed for purposes more directly related to design.

Second, the supersaturation and crystal size may be approximately constant during the growth period, or there may be significant variations in these parameters. In the former case, point values of growth rate are obtained directly; in the latter, point values have to be extracted from the overall system responses. These two cases correspond to the differential and integral techniques respectively, as widely used in chemical reaction engineering.

Third, the measurement defining the growth rate can be obtained from changes in the crystals (for example, increases in their size or mass) or changes in the solution concentration arising from the deposition of solute into the crystal. These two cases, depending on 'solid side' and 'solution side' information respectively, are linked through a mass balance which can be expressed as:

$$-\frac{dw}{dt} = \frac{1}{\rho_L \cdot V_L} \cdot \frac{dM_C}{dt} \qquad (3.1.1)$$

where w is the solution concentration (expressed as a mass fraction), M_C is the total mass of crystals in suspension, ρ_L is the solution density and V_L is the volume of solution in the crystallizing system.

Finally, experiments can be carried out isothermally or non-isothermally. The former is the more usual procedure, although the latter offers the possibility of determining activation energies of crystal growth directly.

3.2 Fixed single crystals

Many different types of information can be obtained by observing a fixed single crystal. For example, the surface structure of a crystal can be observed using a microscope and changes with time recorded. The height, density and lateral velocity, v_s, of moving steps can be measured.

On a larger scale the face growth rates v_{hkl} can be measured. Different faces of a crystal grow at different rates under identical environmental conditions, so these face growth rates do not correlate to overall growth rates in a simple manner (see Section 2.3). In general, high index faces grow faster than low index faces. An accurate assessment of the overall growth kinetics from face growth rates, therefore, must involve a study of several individual face growth rates. Less widely employed is the determination of overall growth rates by weighing individual crystals (gravimetric methods).

Growth rate measurements on fixed single crystals are used in preference to the other methods described in this Chapter for a number of reasons. The investigator may, for methodological reasons, be interested in observing an individual crystal rather than a population of crystals. Features such as surface steps or face growth rates may be of specific interest. Growth rates may need to be studied under conditions where the crystal cannot collide with other objects (the container or other crystals) because of the influence collisions may have on growth kinetics. It may be vital to define the fluid dynamics, or the study of large crystal growth behaviour (larger than about 5 mm) when crystals are difficult to suspend in a fluid might be of interest. All these points can be addressed in fixed single crystal studies in which different types of growth rates can be measured directly. Nevertheless, different methods have to be used for different applications.

The lateral velocities of moving steps and face growth rates can be measured by observing the crystal *in situ* during its growth by means of an optical microscope. Whereas transmission light microscopy is suitable for measuring face growth rates, the measurement of step velocities requires more sophisticated techniques such as interference contrast or phase contrast microscopy. The latter technique is necessary when the dimensions of steps are very small (of the order of a few molecular layers). As a rule, transmission light microscopy is not applicable for the observation of the crystal surface structure because even the clearest crystals obscure the illuminating light to a high degree and thus high quality images with good resolutions cannot be obtained.

Special methods have to be applied to measure the growth rate of very small crystals (attrition fragments and precipitates between about 1 and 500 μm in size). It is difficult to mount crystals of this size by any suitable device and they have to be observed *in situ* while they are lying on a slide.

The definitions of the overall growth rate G and the overall mass disposition rate R_G require observation of the size increase of the crystal as a whole. This can be done by recording the increase of the crystal weight with time.

Description of the experimental equipment necessary to apply these various methods is described in Section 3.2.4.

3.2.1 Solubility, supersaturation and the creation of supersaturation

Accurate adjustment and control of the supersaturation is a prerequisite for the successful measurement of growth rate kinetics. The creation of this super-saturation influences the layout of the measuring equipment and so this topic is discussed first in this section. Only binary solutions are considered.

In a binary solution at isobaric conditions the solubility of the solute in the solvent depends only on the temperature. In many applications this temperature dependence can be used for the creation and control of the supersaturation. However, when the temperature dependence of the solution is not very strong or when the solubility of the component to be crystallized is very low, cooling of the solution does not create a sufficiently high supersaturation. Other methods then have to be applied.

In Table 3.2.1 (see page 54), the solubilities of some substances in aqueous solutions are given for $20°C$ and $30°C$ according to dc^*/dT-values. Several of the substances are very soluble ($c^* = 0.1$ to $10 \, \text{mol dm}^{-3}$) while others are only sparingly soluble. Some of the very soluble substances have a steep temperature dependence, dc^*/dT, but others do not.

As a rule, industrial crystallization processes are carried out at super-saturations Δc between 10^{-3} and $10^{-1} \, \text{kmol m}^{-3}$ [28]. Thus, it can easily be seen that only reasonably soluble substances ($c^* = 0.1$ to $10 \, \text{kmol m}^{-3}$) with a temperature dependence of solubility $dc^*/dT > 10^{-2} \, \text{kmol} \, (\text{m}^{-3} \, \text{K})^{-1}$ can conveniently be studied using a cooling technique. When the temperature dependence is very high (for example Na_2SO_4, $CaCl_2$, KNO_3) temperature control has to be very accurate. On the other hand, low temperature dependence ($CaSO_4$, NaCl) requires extreme temperature variations and for such substances the supersaturation required for the measurement of crystal growth rates has to be achieved by either evaporation or reaction. An alternative technique, drowning-out, will not be considered here because it requires the addition of a third component, the drowning-out agent, to the binary solution.

The remainder of this Section is concerned with the generation of super-saturation in growth cells which form part of a loop apparatus through which the solution circulates. As will be seen in Section 3.2.4, fixed single crystals are

Table 3.2.1 Solubilities of some aqueous systems at 20°C and 30°C[7]

System	$\tilde{M}$ kg kmol^{-1}	W_{20} kg kg^{-1}	W_{30} kg kg^{-1}	c_{20}^* kmol m^{-3}	c_{30}^* kmol m^{-3}	dc^*/dT kmol m^{-3} K^{-1}
Na_2SO_4	142	0.194	0.408	1.315	2.632	$131.7 \cdot 10^{-3}$
$CaCl_2$	110	0.745	1.020	5.783	6.839	$106.6 \cdot 10^{-3}$
KNO_3	101	0.316	0.458	2.782	3.794	$101.2 \cdot 10^{-3}$
$KAl(SO_4)_2$	258	0.059	0.084	0.227	0.321	$94.0 \cdot 10^{-3}$
Urea	76	0.105	0.135	1.437	1.799	$36.2 \cdot 10^{-3}$
KCl	75	0.34	0.37	3.998	4.284	$28.6 \cdot 10^{-3}$
KH_2PO_4	136	0.226	0.277	1.511	1.784	$27.3 \cdot 10^{-3}$
$CuSO_4$	159	0.207	0.250	1.294	1.547	$25.3 \cdot 10^{-3}$
$MgSO_4$	120	0.355	0.408	1.564	1.784	$22.0 \cdot 10^{-3}$
K_2CO_3	138	1.10	1.14	5.917	6.018	$10.1 \cdot 10^{-3}$
$(NH_4)_2SO_4$	132	0.73	0.75	3.356	3.441	$5.9 \cdot 10^{-3}$
Sucrose	342	2.04	2.19	2.590	2.649	$5.9 \cdot 10^{-3}$
NaCl	58	0.360	0.363	5.439	5.454	$1.5 \cdot 10^{-3}$
$CaSO_4$	136	0.0019	0.0020	0.0139	0.0146	$0.07 \cdot 10^{-3}$

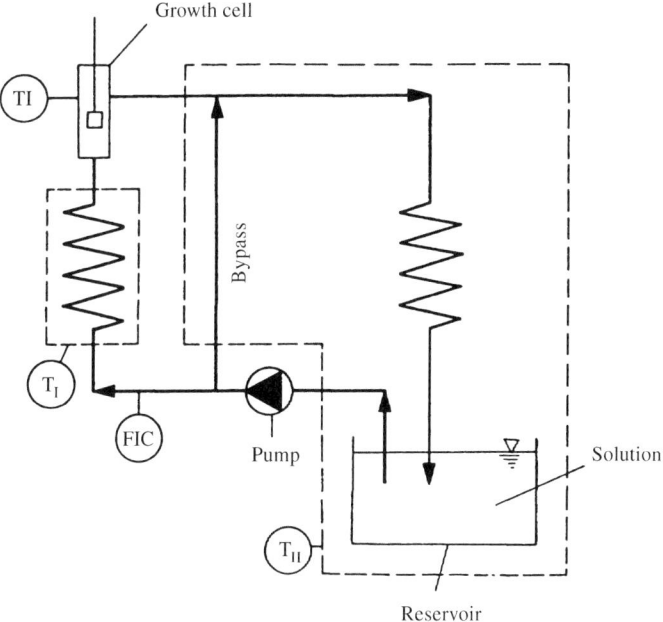

Figure 3.2.1   Loop apparatus in which the solution is to be supersaturated by means of a temperature change

not only studied and observed in such growth cells; stirred vessels and other equipment can be used. Nevertheless, the rules for the creation of super-saturation outlined in this section do not change.

Thermostat T_{II} maintains the main portion of the circulating solution well above saturation, whereas the portion of the solution which passes through the growth cell is supersaturated by reducing its temperature in thermostat T_I. The arrangement illustrated allows rather high supersaturations to be maintained in the growth cell without danger of encrustation or nucleation throughout the solution because the supersaturated solution is mixed with the undersaturated bypass stream immediately following the growth cell. As a rule, the dependence of crystal growth rate on supersaturation should be measured at constant temperature. Therefore, the concentration in the reservoir has to be adjusted by addition of solute or solvent to the solution.

Several authors[29,30] report successful control of supersaturation using a saturator vessel as described by Walker and Kohmann[31]. Figure 3.2.2 (see page 56) depicts the arrangement. The solution circulates through any kind of equipment (here vessel A) in which the crystal is grown at temperature ϑ_1. The concentration of the solution is kept at the desired level by means of the

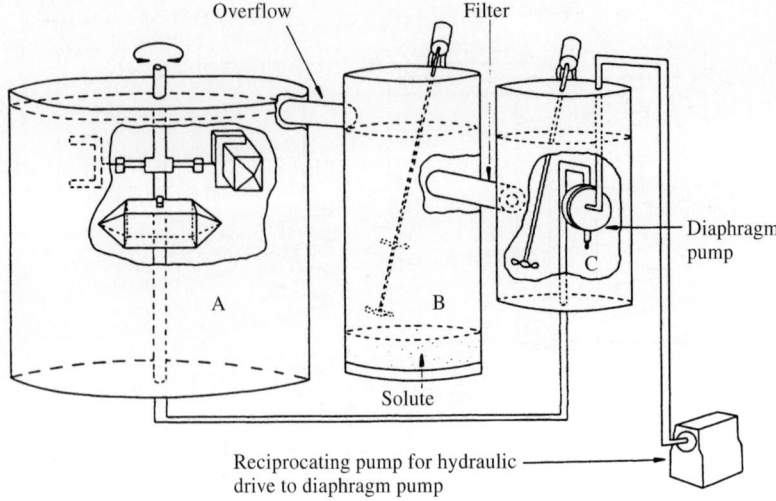

Figure 3.2.2 Schematic diagram of the arrangement of Walker and Kohmann[31]

saturator vessel B, which is operated at temperature $\vartheta_{IV} > \vartheta_I$. The super-saturation σ created by this arrangement is:

$$\sigma = \frac{\Delta c}{c^*} = \frac{(dc^*/d\vartheta)(\vartheta_{IV} - \vartheta_I)}{c^*(\vartheta_I)} \qquad (3.2.1)$$

The fines dissolving vessel C is operated at temperature $\vartheta_{III} > \vartheta_{IV}$, in order to dissolve all fine particles.

In cases where temperature differences cannot provide a sufficiently high supersaturation, reactants or solvent have to be fed to the circulating solution or solvent has to be evaporated from the solution. One temperature level for all the circulating solution is required.

An example for evaporative operation is given in Figure 3.2.3 (see page 57). Solution circulates from the reservoir through the growth cell and back to the reservoir. Solvent is evaporated in the external evaporator and a weighed amount of condensed solvent extracted from the solution. Excess condensate is returned to the reservoir.

Figure 3.2.4 illustrates a loop apparatus in which the supersaturation is created by chemical reaction (see page 57). The principle of operation is very simple with two streams of reactants being fed to the reservoir.

Whereas in the first case high supersaturations can be maintained for considerable lengths of time without danger of massive encrustation or nucleation, in the last two cases high supersaturations eventually lead

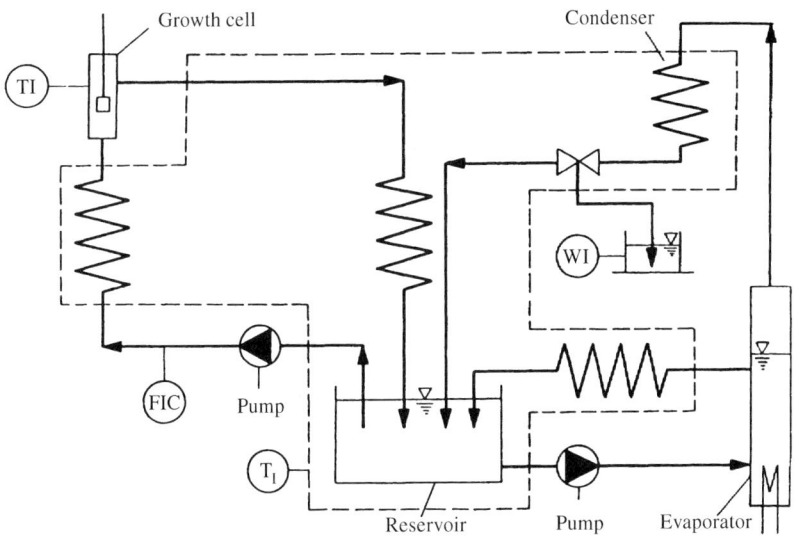

Figure 3.2.3 Loop apparatus with growth cell for evaporation method

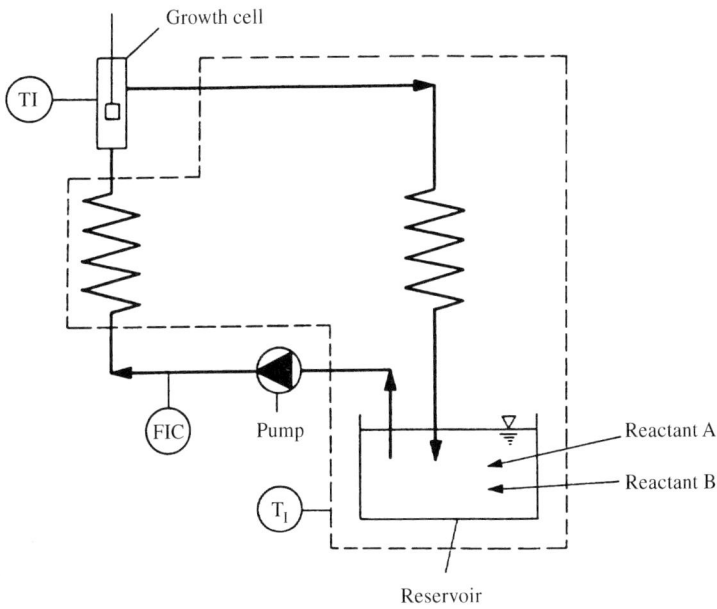

Figure 3.2.4 Loop apparatus with growth cell for reactant feed method

57

to nucleation and solids deposition throughout the entire loop. Solute concentration in the loop is measured using one of the methods described in Section 2.1.

3.2.2 Fluid dynamics

During growth solute has to be transported to, and the solvent transported away from the crystal surface. Fluid flow obviously influences these transport phenomena and has to be well controlled and defined if a complete description of the growth process is required.

Crystal growth processes are considered to consist of two consecutive steps, the diffusion step which accounts for the transport of the solute and solvent to and away from the crystal, and the surface integration step which accounts for surface diffusion and final immobilization of the solute unit in the crystal surface. In Section 2.4 the basic equations for the diffusion and integration steps have been given as well as the relations which combine these steps to the resulting mass deposition rate R_G. Only when the growth of a crystalline substance is solely integration controlled is the fluid dynamics of no importance to mass transfer. In all other cases fluid dynamics does influence the growth rate.

Fluid motion can be present due to forced or natural convection. Because any growing crystal whose growth is not solely integration controlled develops a concentration field adjacent to the crystal and because concentration gradients induce density gradients, the possibility of fluid motion due to natural convection effects has always to be kept in mind. Fluid motion past a growing crystal due to natural convection is very complex and as a rule crystal shape and orientation with respect to the gravitational acceleration have to be considered. Heat effects due to latent heat of crystallization may also sometimes be important. Only high solution viscosities (for example, the sucrose–water system) or low gravitational acceleration may prevent natural convective effects.

Consequently, crystals which grow under diffusion controlled conditions in low viscosity systems should not be studied without forced convection overwhelming any hard-to-define natural convection effects. A crystal face oriented, and preferably perpendicular, to the stream lines of controlled forced convective flow grows under well defined fluid dynamic conditions. In the case of turbulent flow, an intensity of turbulence greater than zero adds to the mass transfer. It should also be noted that fluid dynamics not only contribute to mass transfer but also act in a very complex manner on the formation of liquid inclusions[29].

3.2.3 Layout of equipment

In principle, three different methods for measuring the growth kinetics of fixed single crystals are described in the literature. If L is some characteristic dimension of the crystal, then these methods can be categorized as being applicable to large single crystals with L > 5 mm, small single crystals with 1 mm < L < 5 mm, with single crystals which are too small to be fixed by some device (L < 0.5 mm).

In Sections a to c following, the equipment layout for these three is described while the set-up for microtopographic investigation of crystal surfaces (morphology, velocity and height of moving steps) is explained in Section d.

a) Large crystals (L > 5 mm)

Observing the fluid dynamic recommendations given in Section 3.2.3, fixed single crystals should be exposed to well defined forced convective (laminar or turbulent). This can be achieved best by placing the crystal at the downstream end of a tube of a known diameter and length. The tube diameter should be at least three times the crystal diameter. Thus, a 5 mm crystal requires a 15 mm diameter tube. If mean forced convection velocities of say $0.3 \, \mathrm{m \, s^{-1}}$ are to be achieved a flow rate of $3.2 \, \mathrm{l \, min^{-1}}$ is then required. At the laboratory scale these, and even much higher flow rates for bigger crystals, are not easy to handle with respect to temperature and concentration control. Consequently, other techniques have to be applied for crystals larger than about 5 mm.

Two techniques have been developed. Either the crystal is placed in an agitated solution or the crystal itself is fixed on a rotor and turned in the solution. Although both methods are similar, the fluid dynamics are different. The flow past the crystal is likely to be more turbulent and multidirectional in an agitated solution than in the latter technique. On the other hand, access to the crystal, for instance for weighing, is much easier if the crystal is not mounted on a rather mechanically complicated rotor.

Figure 3.2.5 (see page 60) depicts the layout of the apparatus used by Janssen-van Rosmalen et al.[29] They used a three vessel system for super-saturation control as described in Section 3.2.1. Vessel I is the saturator, vessel II the fines dissolver and vessel III the crystallizer with four crystals fixed to a single rotor. Vessel III is equipped with baffles in order to prevent the bulk of the solution from rotating.

Figure 3.2.6 (see page 60) depicts the layout of the apparatus used by Smythe[32]. The jacketed baffled beaker, A, of approximately 100 ml capacity contains the supersaturated solution which is agitated by a two blade impeller. The weight of the crystal, which is held by stainless steel crystal tongs, is determined by a torsion balance.

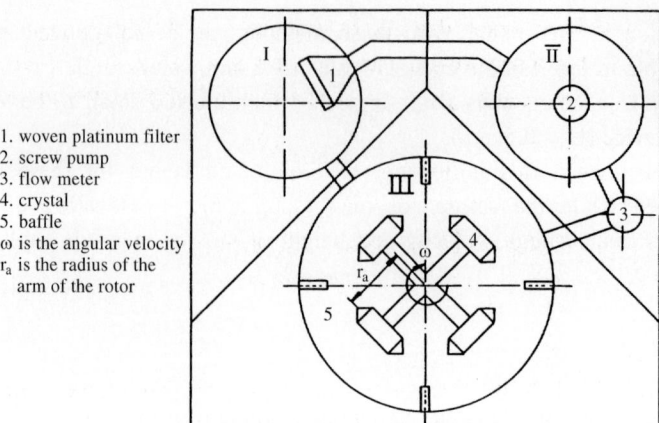

1. woven platinum filter
2. screw pump
3. flow meter
4. crystal
5. baffle
ω is the angular velocity
r_a is the radius of the
   arm of the rotor

Figure 3.2.5 Crystal growth equipment according to Janssen-van Rosmalen *et al*[29].

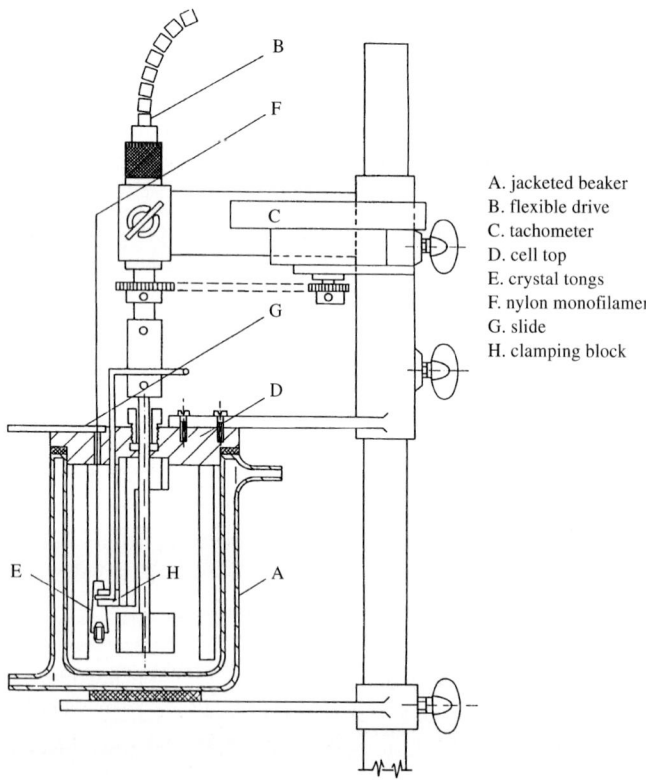

A. jacketed beaker
B. flexible drive
C. tachometer
D. cell top
E. crystal tongs
F. nylon monofilament
G. slide
H. clamping block

Figure 3.2.6 Apparatus according to Smythe[32]

Several authors report that the crystal may be weighed by means of a balance[32-34]. From the weight increase with time $\Delta M/\Delta t$ the overall mass deposition rate R_G can be calculated according to (compare Section 3.5):

$$R_G = \frac{\Delta M_C}{A_C \cdot \Delta t} \quad \text{with} \quad A_C = \beta \cdot \left(\frac{M_C}{\rho_C \cdot \alpha}\right)^{3/2} \qquad (3.2.2)$$

In order to obtain the weight increase ΔM_C or weight M_C of the crystal, the balance reading has to be corrected for buoyancy.

b) Small crystals (5 mm > L > 1 mm)

For small crystals in the size range from 1 to 5 mm, growth cells are widely used. They permit the crystal faces to be oriented with respect to the flow and thus reproducible growth conditions to be achieved. As a rule growth cells are used for the measurement of face growth rates v_{hkl} or for the observation of growth steps on the surfaces. Figure 3.2.7 depicts the growth cell used by

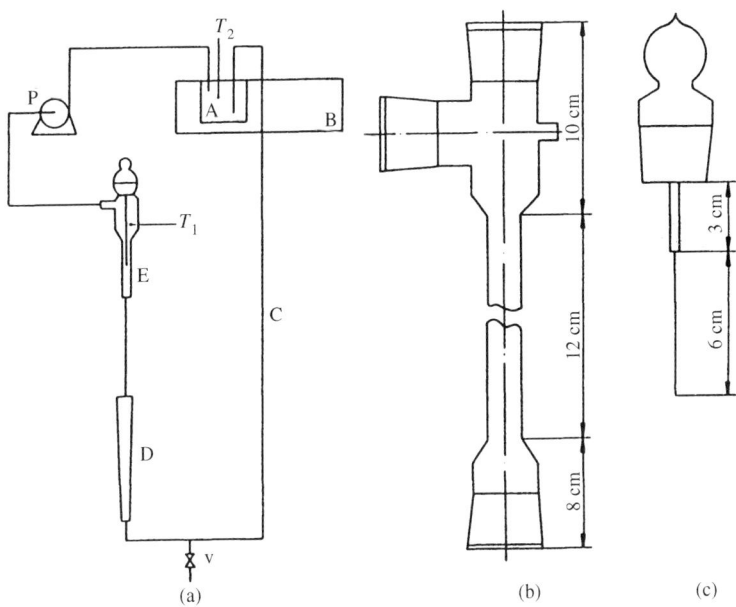

(a) Loop apparatus, A reservoir, B thermostated bath, D rotameter, E growth cell, P pump, T_1, T_2 thermometer
(b) Growth cell made from pyrex glass. Centre section ranges from 5 to 10 mm in diameter
(c) Crystal holder. The crystal is cemented with a nitro-cellulose based adhesive to the end of a 1mm diameter tungsten wire

Figure 3.2.7 Growth cell according to Mullin and Amatavivadhana[35]

Mullin and Amatavivadhana[35]. Several authors have worked with this or similar growth cells[36–41].

The minimum crystal size is limited to about 1 mm by the mounting method. No method which allows smaller crystals to be fixed without disturbing the fluid dynamics near the crystal has been developed yet.

Special attention has to be paid to the optical set-up. As a rule, microscopes are used to follow the advance of the faces *in situ*. Such microscopes must have a long working distance and high depth of field. The optical quality of the windows through which the crystal is observed is crucial for good image resolution. It is best to use a cold light source and the light should be monochromatic by means of a filter to prevent prismatic effects. Figure 3.2.8 depicts an improved growth cell. The crystal is here observed through microscope slides which are attached to the cell.

Fluid dynamic considerations require that the crystal be placed at the centre line of the tube leading into the growth cell. It is known that in pipe flow a length to diameter ratio of about $0.03 \cdot Re$ is required in order to obtain a fully developed laminar flow[42]. For such a flow profile the fluid velocity at the centre line is twice the mean velocity. In growth cells, where the tube is usually shorter than specified by the above criterion, the laminar profile is not fully developed at the crystal position and the intensity of turbulence, which most probably cannot be maintained to be zero at the tube inlet may not have vanished totally at the crystal position.

c) Very small crystals (500 μm > L > 1 μm)

An important area in industrial crystallization concerns the generation and growth of secondary nuclei. For the purpose of observing such nuclei, the

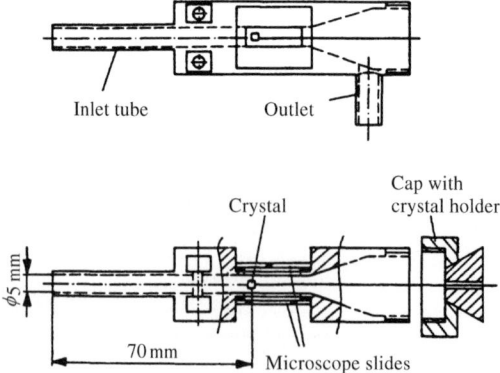

Figure 3.2.8   Improved growth cell with microscope slides

growth cell shown in Figure 3.2.9 was introduced by Garside and Larson[43]. Attrition fragments produced by contacting the parent crystal (2) fall onto the central cover glass and can be observed during their growth by means of transmission light microscopy. In order to automate the experiment Wang and Mersmann[44] recommend the use of a video microscope system. A more detailed observation of the formation as well as the initial growing of attrition fragments then becomes possible.

Although not designed for the purpose, the growth cell may also be used for observing the growth behaviour of any kind of very small crystals. Garside et al.[45], Berglund et al.[46], and Ramanarayaran et al.[47], used such a cell to determine the growth behaviour and size distribution of contact nuclei of various substances.

Because no forced convective fluid motion can be imposed, experiments carried out in this way have certain drawbacks. For example, desupersaturation of the solution cannot be prevented and this takes place in the solution as a whole as well as in the direct environment of the growing crystals.

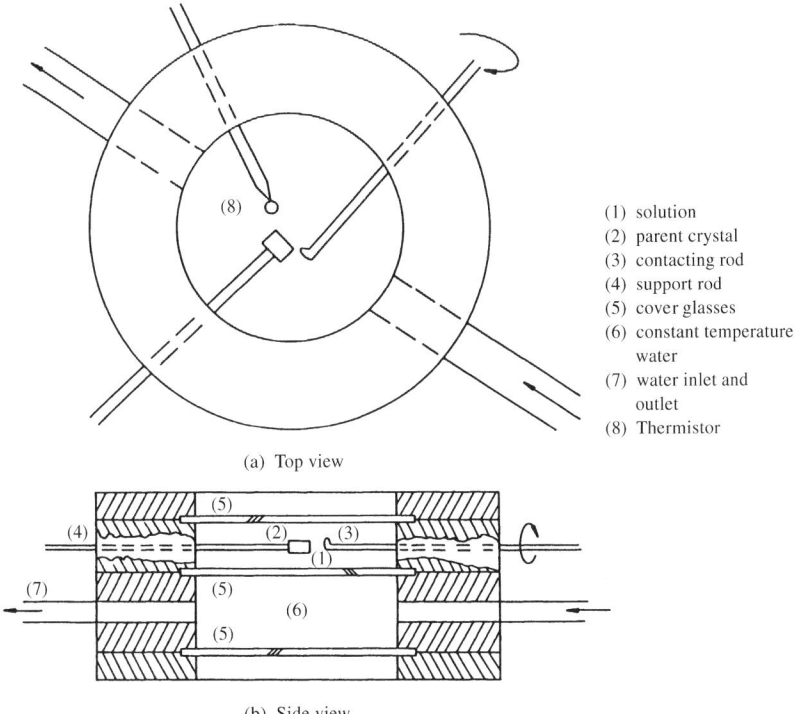

(1) solution
(2) parent crystal
(3) contacting rod
(4) support rod
(5) cover glasses
(6) constant temperature water
(7) water inlet and outlet
(8) Thermistor

(a) Top view

(b) Side view

Figure 3.2.9 Schematic diagram of the cell used by Garside and Larson[43]

The concentration field of one particle is distorted because it lies on the cover glass and the concentration fields of two particles lying close together may interfere with one another. Natural convection may also be present. All of these drawbacks are of minor importance when the crystals being observed are growing in the integration controlled regime.

d) Investigation of crystal surfaces

The investigation of crystal surfaces is mostly of scientific rather than technical interest. Nevertheless, we briefly review this topic to give an idea of what is possible in fixed single crystal growth measurements.

For microtopographical purposes reflection microscopy is applied. The main objective is to reveal growth steps and other kinds of undulations on the crystal surface. This implies that the optical microscope must be capable of detecting both extremely small height differences, as low as one atomic layer, and very small inclinations, down to about 0.05 degrees. The only way to achieve this is to convert height differences, into phase differences, which are then transformed into intensity differences. This transformation can be carried out by either Fourier or interferometric techniques.

Fourier techniques consist of either dark field illumination, pencil illumination, the Schlieren method, or phase contrast microscopy. Interference microscopy covers interferometry, by which quantitative information can be obtained, or interference contrast microscopy, by which phase differences are translated into intensity or colour differences[152].

In the last decade fascinating progress has been made with these techniques and many papers have been published by Van Enckevort and co-workers (summarized in Van Enckevort[152]) and Tsukamoto and his co-workers[49]. These researchers have even made *in situ* observations of monomolecular growth steps on crystals growing in solution or melt.

Kämmer *et al.*[50] used a scanning tunnelling microscope to investigate the surface morphology of KNO_3 crystals. With this measuring device they were able to observe growth steps of a height of approximately 8 nm.

3.2.4 Problems

Many of the problems which may occur with certain compounds or with the operation of the growth apparatus have already been mentioned[146,151,152]. Specific problems concerning impurities in the solution and the interpretation of measured data will be addressed here.

It is known that impurities (undesired) or additives (desired) may influence the face growth rate of a crystal, even when present at concentrations in the ppm range. For example Seifert[51] and Tengler *et al.*[52] could not grow KCl crystals from solution until recrystallization removed most of the lead ions, which supposedly retarded growth to zero. Multi particle systems are not so sensitive to impurities, because the adsorption surface is large and the resulting surface impurity concentration is low. A single crystal in solution, by contrast, has a comparatively small surface area and thus, the surface impurity concentration may be very high. As a result care should be taken that no impurities are introduced into the solution by the apparatus itself. Only materials such as glass, silicon and Teflon should be used.

3.2.5 Example

Assume that the growth kinetics of $(NH_4)_2SO_4$ in aqueous solution have to be measured at a temperature of 29°C.

First, the solubility curve of $(NH_4)_2SO_4$ has to be known. It can be taken from Mullin[7] for example, and is reproduced in the following Table 3.2.2 and in Figure 3.2.10 (see page 66):

In the range of 30°C the solubility curve can be linearized to:

$$w = 0.413 \frac{kg_{salt}}{kg_{solution}} + 8.5 \cdot 10^{-4} \frac{kg_{salt}}{kg_{solution}} \degree C^{-1} \cdot \vartheta \qquad (3.2.3)$$

For a reliable determination of the supersaturation of the solution it is necessary to measure its temperature and concentration. As was shown in Section 2.1.2, different methods (measurement of density, refractive index or electrical conductivity, or drying out of a sample) can be applied for concentration measurements. In this case density measurement is chosen.

The dependence of the density of the solution on its saturation temperature is reported in literature[7]. Figure 3.2.11 (see page 66) shows that these data are not reliable enough for the relatively small subcoolings that will have to be controlled (as can be seen later).

The apparatus depicted in Figure 3.2.1 (see page 55) is suitable for the measurement of the growth kinetics of this system. Solution of approximately known saturation temperature $\vartheta^* \approx 31 \ldots 33\degree C$ is charged into the apparatus.

Table 3.2.2 Solubility of $(NH_4)_2SO_4$ in water

ϑ	°C	0	10	20	30	40	60	80	100
W	$\frac{kg_{Salt}}{kg_{Water}}$	0.710	0.730	0.754	0.780	0.810	0.880	0.953	1.033
w	$\frac{kg_{Salt}}{kg_{Solution}}$	0.415	0.422	0.430	0.438	0.447	0.468	0.488	0.508

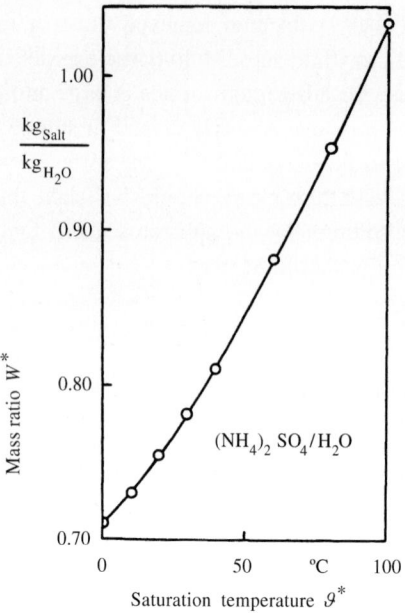

Figure 3.2.10   Solubility of $(NH_4)_2SO_4$ in water

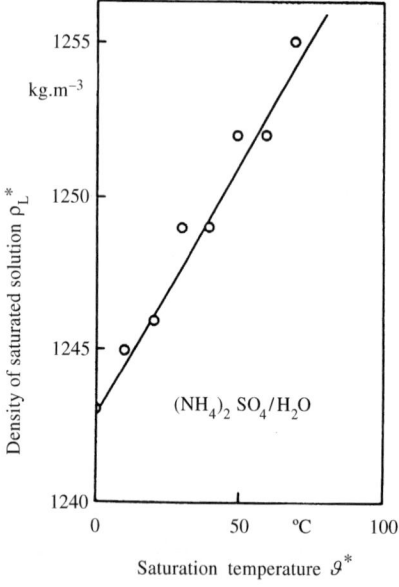

Figure 3.2.11 Temperature dependence of the density of saturated aqueous $(NH_4)_2SO_4$ in water

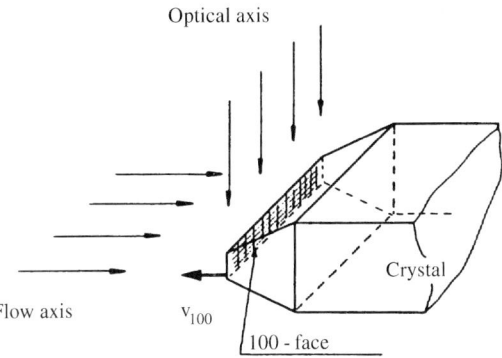

Figure 3.2.12 Orientation of a crystal in the growth cell

The equipment is started up and kept running at $\vartheta > \vartheta^*$ until temperature stability is achieved and TI indicates a temperature of $\vartheta > \vartheta^*$. Before inserting the crystal the solution concentration has to be measured exactly. After insertion, the crystal begins to dissolve because $\vartheta > \vartheta^*$. After a few minutes the solution temperature in the growth cell should be cooled down to the measuring temperature $\vartheta = 29°C$. The crystal will then start to grow and sharp edges and flat faces should appear.

It is important to have the desired crystal face properly oriented to the flow as well as to the optical axis of the microscope as shown in Figure 3.2.12.

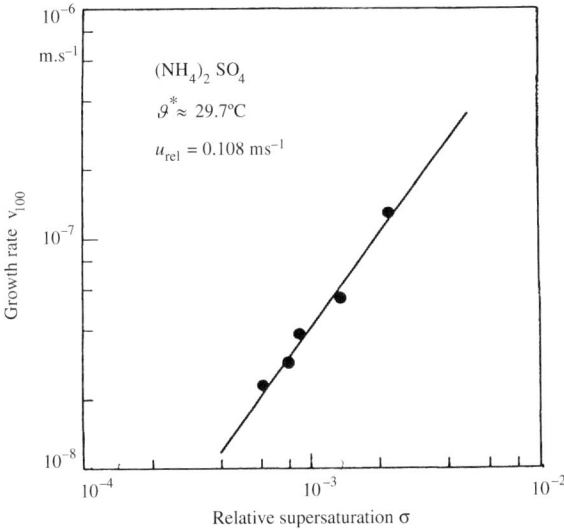

Figure 3.2.13 Growth of 100-face versus relative supersaturation

Table 3.2.3 Example set of experimental data for the measurement of growth kinetics of the 100-face of $(NH_4)_2SO_4$ at $\simeq 29.7°C$ and a solution velocity of $u_{rel} = 0.108\,ms^{-1}$ [40]

No	w kg/kg solution	$\vartheta*$ °C	σ –	ΔL_{100} m	Δt s	v_{100} ms^{-1}
1	0.43793	29.3	$6.39 \cdot 10^{-4}$	$1.65 \cdot 10^{-4}$	7200	$0.23 \cdot 10^{-7}$
2	0.43800	29.4	$8.06 \cdot 10^{-4}$	$2.09 \cdot 10^{-4}$	7200	$0.29 \cdot 10^{-7}$
3	0.43805	29.5	$9.13 \cdot 10^{-4}$	$2.05 \cdot 10^{-4}$	5400	$0.38 \cdot 10^{-7}$
4	0.43825	29.7	$1.37 \cdot 10^{-3}$	$1.98 \cdot 10^{-4}$	3600	$0.55 \cdot 10^{-7}$
5	0.43863	30.1	$2.23 \cdot 10^{-3}$	$2.32 \cdot 10^{-4}$	1800	$1.29 \cdot 10^{-7}$

The growth velocity of the desired face is then measured by the displacement ΔL_{100} of the (100)-face per unit time Δt.

In Table 3.2.3 an example set of data is presented[40]. The different saturation temperatures have been adjusted by adding either water or $(NH_4)_2SO_4$ to the solution. The solution velocity was kept constant at $u_{rel} = 0.108\,m\,s^{-1}$ while the measuring temperature was $\vartheta = 29.0°C$ ($\approx w^* = 0.4379\,kg$ per kg solution).

By regression analysis, from these data the following equation law can be derived (compare Figure 3.2.13):

$$v_{100} = 4.85 \cdot 10^{-4} \cdot \sigma^{1.36} [m\,s^{-1}] \qquad (3.2.4)$$

3.3 Rotating disc

3.3.1 Introduction

In measuring growth kinetics, it is frequently important to be able to separate the roles of diffusion and integration, particularly if fundamental information on the surface process is required or if the results have to be extrapolated to other temperatures or relative crystal/solution velocities. In such circumstances the experimental technique must be such that the diffusion or mass transfer step is well characterized. The rotating disc system is one technique that enables such information to be obtained.

The rotating disc has been widely used in the field of electrochemistry[53] and there are a small number of crystal growth studies that have employed the technique[54–57].

3.3.2 Principles

The rotating disc geometry is very simple. A disc is rotated in a horizontal plane about its axis with the angular velocity $\omega = 2\pi s$. This induces flow towards the disc and then radially outwards across the disc surface. With the assumption that the disc is in an infinite expanse of fluid and that edge effects are negligible, the fluid flow equations can be solved to give the hydrodynamic boundary layer thickness[58]:

$$\delta_H = 2.8 \cdot \left(\frac{\nu_L}{\omega}\right)^{1/2} \tag{3.3.1}$$

Note that the boundary layer thickness is independent of position on the disc surface and so the disc is 'uniformly accessible' to solute.

The corresponding mass transfer case can be solved and, when written in terms of a mass transfer coefficient, results in[59]:

$$k_d = 0.6205 \cdot D^{2/3} \cdot \frac{\omega^{1/2}}{\nu_L^{1/6}} \cdot f(Sc) \tag{3.3.2}$$

The function $f(Sc)$ is given by Newman[59] as:

$$f(Sc) = 1 + 0.2980 \cdot Sc^{-1/3} + 0.1451 \cdot Sc^{-2/3} \tag{3.3.3}$$

This expression plotted in Figure 3.3.1 $f(Sc)$ approaches unity at large values of Sc and at $Sc = 10^3$, $f(Sc) = 1.031$. In liquid solutions, therefore, $f(Sc)$ can be neglected if an error of up to 4% is acceptable.

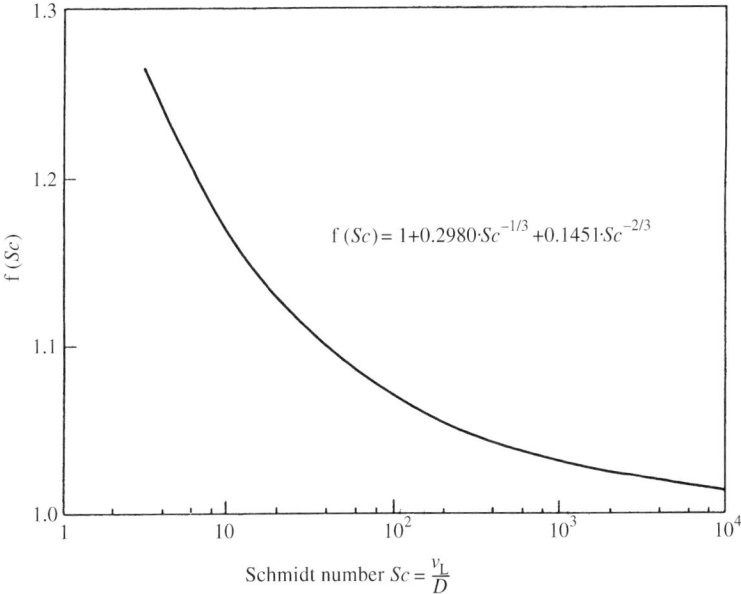

Figure 3.3.1 Schmidt number correction factor

If the diffusion coefficient for the system is known, Equation (3.3.2) can be used to calculate k_d directly. Alternatively dissolution experiments can be used to determine the mass transfer coefficient through the equation:

$$R_D = \rho_L \cdot k_d \cdot \ln\left(\frac{1 - w_\infty}{1 - w^*}\right) = \rho_L \cdot k_d \cdot \ln(1 + B) \tag{3.3.4}$$

where driving force B is defined by:

$$B = \frac{w^* - w_\infty}{1 - w^*} \tag{3.3.5}$$

If $B \ll 1$, $\ln(1 + B) \approx B$ and so Equation (3.3.4) can be written:

$$R_D = \rho_L \cdot k_d \cdot \left(\frac{w^* - w_\infty}{1 - w^*}\right) = k_d \cdot \left(\frac{C^* - C_\infty}{1 - w^*}\right) \tag{3.3.6}$$

The resulting value of k_d can then be used to describe the diffusion step during crystal growth as outlined in Section 2.4.1.

The rotating disc technique can be used to measure the growth rate either of individual crystal faces[54] or of multi-crystal systems[55-57]. In the former case one crystal is mounted in the rotating disc with a specific crystallographic face exposed, in the latter case many crystals are compressed into a recess in the disc surface, their random orientation exposing the crystallographic faces present on the crystals.

3.3.3 Equipment

The equipment comprises two main parts (a) the rotating disc and (b) the motor assembly and solution reservoir. A schematic outline is shown in Figure 3.3.2.

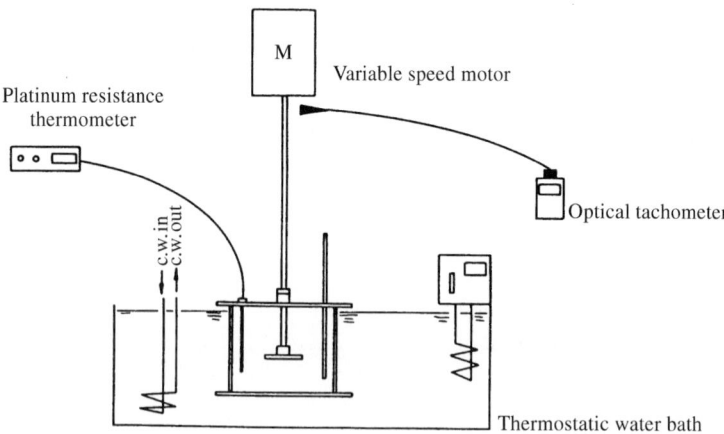

Figure 3.3.2 Schematic diagram of rotating disc experimental arrangement

(a) The rotating disc

The disc may be made from a number of materials to suit the specific solution with which it will be in contact. Aluminium, stainless steel and perspex have been used. The disc is mounted on a shaft which is connected to a motor and care must be taken to see that the disc is horizontal and that no eccentricity occurs. A recess is cut into the disc into which the crystal(s) is mounted, either by securing a single crystal or compressing a large number of crystals.

A number of criteria must be satisfied to ensure that the assumptions made in the derivation of the rotating disc equations hold. These relate to the disc dimensions relative to the boundary layer thickness and to other operating conditions. To illustrate these points let us assume that the solution has similar properties to water; specific physical properties are thus taken to be $v_L = 10^{-6}\,\mathrm{m^2\,s^{-1}}$ and $D = 10^{-9}\,\mathrm{m^2\,s^{-1}}$. Hence $Sc = v_L/D = 10^3$ and the ratio of the hydrodynamic to diffusion boundary layer thickness, $\delta_H/\delta_M \approx (Sc)^{1/3} = 10$. The hydrodynamic boundary layer is an order of magnitude thicker than the diffusion boundary layer.

Equation (3.3.1) defines the value of δ_H. For disc rotation speeds between 50 and 1000 rpm (5.2 and 105 rad s^{-1} respectively) δ_H is thus between about 1.2 and 0.27 mm. The corresponding values of δ_M are about 120 and 27 μm.

Various criteria can now be examined:

- For edge effects to be negligible the radius, r, must be greater than δ_H. This is satisfied so long as the disc radius is greater than about 10 mm.
- The assumption of an infinite expanse of fluid will be satisfied so long as the clearance between the disc and the solution reservoir walls and air/solution interface are much greater than δ_H. This should be easy to satisfy.
- The radius of the central 'active portion' of the disc from which mass transfer takes place must be much larger than the diffusion boundary layer, δ_M. For high Schmidt numbers this does not present any difficulty.
- Flow over the disc must be laminar. The disc Reynolds number at which turbulent flow starts depends on the surface finish of the disc. A maximum value of $Re_{disk} = r^2 \cdot \omega/v_L = 2 \cdot 10^5$ will ensure laminar flow. This value could be exceeded with large discs and high rotation speeds. For example with $r = 50$ mm, $\omega = 105$ rad s^{-1} ($s = 16.7\,\mathrm{s^{-1}}$) and $v_L = 10^{-6}\,\mathrm{m^2\,s^{-1}}$, $Re_{disk} = 2.6 \cdot 10^5$.
- At low values of Re_{disk} natural convection can be significant and so Re_{disk} should exceed a minimum value of about 10^{2} [60]. For $r = 25$ mm, $\omega = 6.3$ rad s^{-1} ($s = 1\,\mathrm{s^{-1}}$) and $v_L = 10^{-6}\,\mathrm{m^2\,s^{-1}}$, $Re = 390$.
- The disc surface should be 'smooth'. This criterion is met if the height of the roughness elements are less than the boundary layer thickness. Roughness is most likely to arise from the growth of individual crystals in a polycrystalline

disc. It is perhaps unlikely to influence the hydrodynamic boundary layer but it is possible that the diffusion boundary layer, δ_M, will be affected.

The above discussion indicates that most of the necessary criteria can be met. It should be noted however that if solutions of high viscosity are used the boundary layer thickness is greater, and some of these criteria become more difficult to satisfy. For example an aqueous solution of sucrose, saturated at 35°C has the physical properties $v_L = 1.27 \cdot 10^{-4} \, m^2 \, s^{-1}$ [61] and $D = 8.6 \cdot 10^{-11} \, m^2 \, s^{-1}$ [62]. Thus $Sc = 1.5 \cdot 10^6$ and $\vartheta_H/\delta_M \approx 110$. For $\omega = 36.7 \, rad \, s^{-1}$ (350 rpm), $\delta_H \approx 5 \, mm$.

Figure 3.3.3 gives the dimensions of a disc that satisfies the criteria set out above. Using a polycrystalline growth surface the growth rate can best be determined from the increase in weight of the whole disc. If the active region grows by 0.1 mm during a growth experiment, the increase of mass for a crystal of density $1500 \, kg \, m^{-3}$ over the active radius of 15 mm is about 0.1 g.

In order to increase the accuracy in determining the increase in disc weight, a removable insert can be placed in the disc recess. This can then be removed with the crystal mass for weighing purposes.

(b) Motor assembly and solution reservoir

The remainder of the equipment is straightforward, (Figure 3.3.2 on page 70). The motor assembly is required to maintain a constant, but variable, speed for periods of time of up to several hours. The rotation speed must be measured.

The solution must be kept at constant temperature and the solution reservoir is best suspended in a thermostated water bath. Ideally the solution temperature should be maintained to within ±0.01 K, although the exact requirement will depend on the temperature coefficient for the system under study. The reservoir itself must be sufficiently large so that clearances between the walls, base and air/solution interface are much greater than δ_H. For $\delta_H \sim 1 \, mm$ a 20 mm clearance is adequate so giving a reservoir of minimum diameter $\sim 65 \, mm$. The

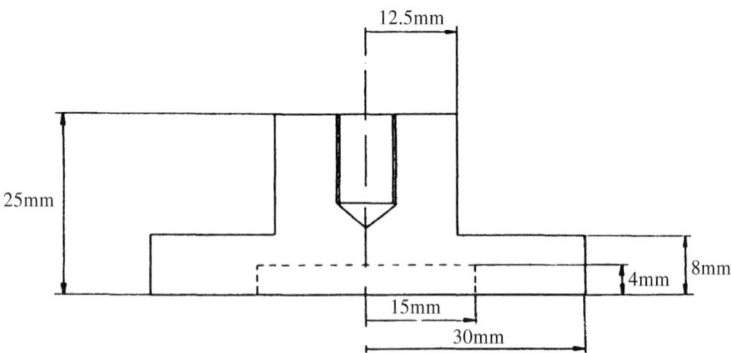

Figure 3.3.3   Details of rotating disc

resulting volume of about $0.5\,\text{dm}^3$ is rather small and in practice a volume of $2-3\,\text{dm}^3$ is more convenient. A typical reservoir dimension would then be about 150 mm diameter $\times$ 120 mm deep. Typical construction details are illustrated in Figure 3.3.4[63]. A glass pipe section is clamped between two end plates and sealed by rubber gaskets and silicon rubber sealant. The upper plate contains a thermometer port and sample ports while the disc shaft is sealed with a teflon insert.

3.3.4 Measurement procedure

(a) Solution preparation

Solution of the required concentration is prepared at a temperature above its saturation point. This solution is filtered and charged into the solution reservoir. A blank disc can be used to gently agitate the solution while a sample is taken for the initial concentration measurement and the required working temperature established.

The 'active' disc containing the crystal(s) is weighed, preheated to about the working temperature and then quickly exchanged with the blank disc. The run is started immediately. It is advisable to try to vary the run length so that the amount of growth on the crystal(s) is about the same in all runs, the run time therefore varying with solution supersaturation. For a polycrystalline disc an increase in weight of between about 0.1 and 0.2 g is convenient.

After growth has proceeded for the estimated time the rotor is stopped, dismantled, the disc removed from the solution and the crystal(s) immediately

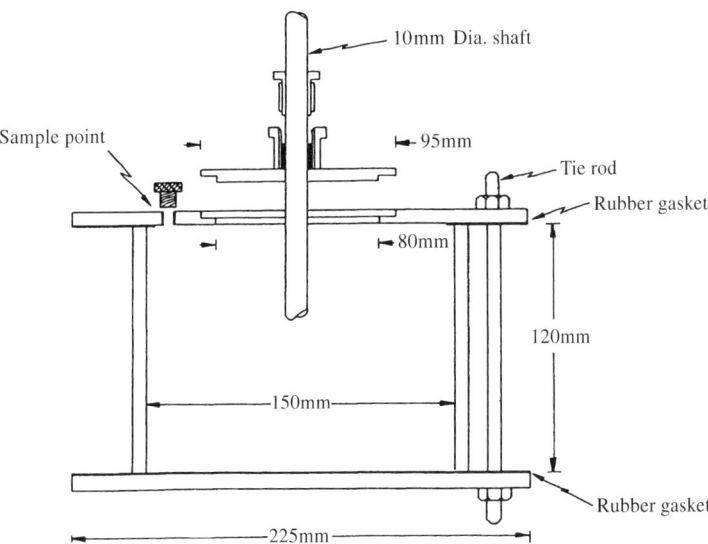

Figure 3.3.4   Construction details of rotating disc reservoir

wiped clean of any excess solution with tissues. It is then unscrewed from the shaft, cleaned, dried and weighed. A final solution sample is also taken for concentration measurement. The disc drying procedure is critical for accurate results and must be developed for the specific system.

(b) *Disc preparation*

The most satisfactory preparation method for a polycrystalline disc will depend on the actual material being crystallized and will have to be evolved by trial and error. The following technique has been found satisfactory for sucrose and should form a good starting point for any further development that may be needed.

A mixture of crystals and concentrated solution (mass ratio of crystals to solution ~ 5:1) is pressed into the disc recess using an hydraulic ram press. The disc is then oven dried. Such discs tend to absorb solution and so a 'conditioning' process is needed to fill the pore structure in the polycrystalline mass. Thus the disc is immersed in a saturated solution for a few minutes, removed, dried and weighed. The process is repeated until constant mass is achieved.

3.3.5 Example of measured data

The results described here refer to growth on a polycrystalline disc of sucrose growing in aqueous solution at $31°C^{63}$.

By inspection of Equation (3.3.6) a plot of dissolution rate R_D against the bulk concentration w_∞ (or C_∞) should be linear with the intercept on the concentration axis representing the saturation concentration w^* (or C^*). This saturation value can thus be determined or a check made on published data. In this case w^* was estimated to be $w^* = 0.6835$ kg sucrose per kg solution. The linearity of the dissolution rate with B is demonstrated in Figure 3.3.5 on page 75, these data referring to a rotation speed $\omega = 26.2$ rad s^{-1} (250 rpm).

From the dissolution rate curve the mass transfer coefficient can be calculated from Equation (3.3.6) and in this case has the value $2.45 \cdot 10^{-6}$ m s^{-1}. Equations (3.3.2) and (3.3.3) then enable the diffusion coefficient to be estimated. In this particular instance the relevant parameters are:

$$\rho_L = 1330 \text{ kg m}^{-3}$$
$$\eta_L = 0.135 \text{ Pa s}$$
$$v_L = \frac{\eta_L}{\rho_L} = 1.015 \cdot 10^{-4} \text{ m}^2 \text{ s}^{-1}$$
$$\omega = 26.2 \text{ rad s}^{-1}$$
$$s = 259 \text{ min}^{-1}$$
$$k_d = 2.45 \cdot 10^{-6} \text{ m s}^{-1}$$

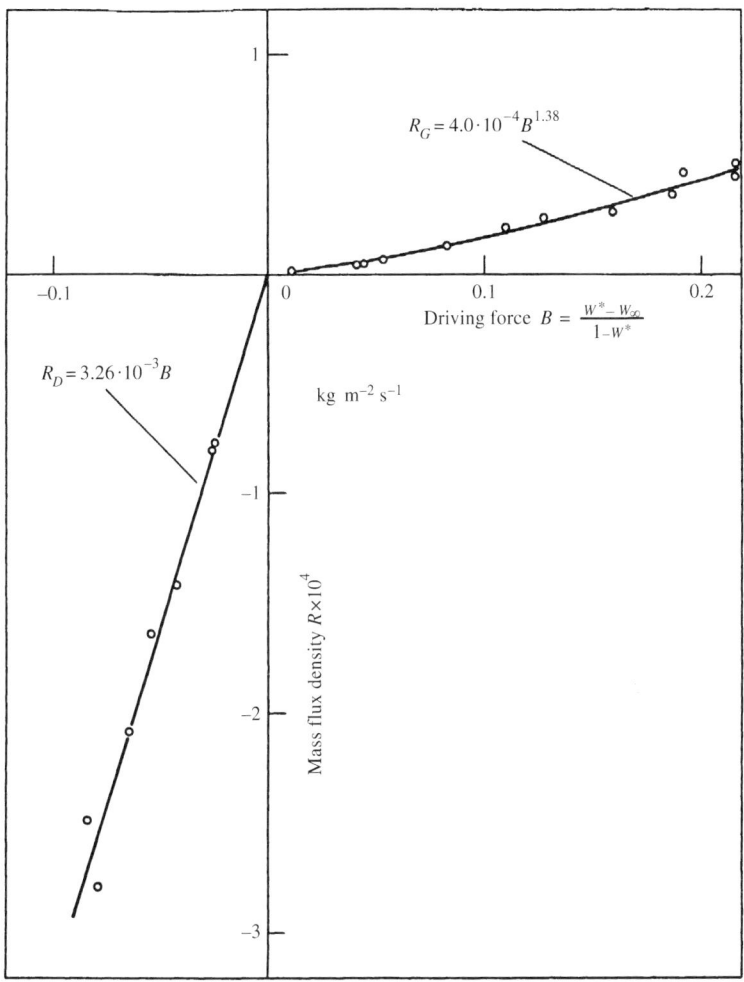

Figure 3.3.5 Overall growth and dissolution rate in pure sucrose solution and 31°C ($\omega = 26.2\,\mathrm{rad\,s^{-1}}$)

Assuming f(Sc) = 1.00:

$$D = \left[\frac{2.45 \cdot 10^{-6} \cdot (1.015 \cdot 10^{-4})^{1/6}}{0.6205 \cdot (26.2)^{1/2}} \right]^{3/2}$$

so

$$D = 6.80 \cdot 10^{-11}\ \mathrm{m^2\,s^{-1}}$$

With this value of D, Sc can be estimated:

$$Sc = \frac{v_L}{D} = \frac{1.015 \cdot 10^{-4}}{6.80 \cdot 10^{-11}} = 1.49 \cdot 10^6$$

Equation (3.3.3) then indicates that $f(Sc) = 1.003$ and so the assumption that $f(Sc) = 1.0$ is clearly justified.

Note that in this particular case the values of $B = (w^* - w_\infty)/(1 - w^*)$ are comparatively large and so the criterion $B \ll 1$ is not strictly satisfied. If the dissolution rates are correlated with $\ln(1 + B)$ rather than B (see Equation (3.3.4)) the resulting value of k_d is $2.58 \cdot 10^{-6}\,\mathrm{m\,s^{-1}}$, giving a diffusion coefficient $D = 7.35 \cdot 10^{-11}\,\mathrm{m^2\,s^{-1}}$.

Interpolation of published diffusivity data obtained using an interferometric technique[62] give a value of $D = 8.1 \cdot 10^{-11}\,\mathrm{m^2\,s^{-1}}$ for the temperature and bulk concentration used in the experiment described above.

Figure 3.3.5 also illustrates overall crystal growth rates of sucrose under similar conditions, here plotted against the B driving force. In this case growth is much slower than dissolution, indicating that the integration step dominates the process. The overall growth kinetics can be correlated by the empirical Equation (2.4.1):

$$R_G = k_G' \cdot (w_\infty - w^*)^g = k_G \cdot (C_\infty - C^*)^g$$

with

$$k_G' = \rho_L^g \cdot k_G$$

which in this case gives $g = 1.48$ and $k_G' = 3.27 \cdot 10^{-3}\,\mathrm{kg\,m^{-2}\,s^{-1}}$.

The surface integration kinetics can be evaluated by determining the solution concentrations at the crystal/solution interface as outlined in Section 2.4.1. For one representative experimental point for example:

$$R_G = 1.38 \cdot 10^{-5}\,\mathrm{kg\,m^{-2}\,s^{-1}} \text{ at } (w_\infty - w^*)$$
$$= 0.02421\,\mathrm{kg}\text{ sucrose per kg solution}$$

$$(w_I - w^*) = (w_\infty - w^*) - R_G \cdot (1 - w_\infty)/\rho_L \cdot k_d$$

For $k_d = 2.45 \cdot 10^{-6}\,\mathrm{m\,s^{-1}}$ and $\rho_L = 1330\,\mathrm{kg\,m^{-3}}$

$$(w_I - w^*) = 0.02297\,\mathrm{kg}\text{ sucrose per kg solution.}$$

The major part of the overall driving force is therefore associated with the integration step, since only a small driving force, $0.00124\,\mathrm{kg}$ sucrose per kg solution in this case, is required to drive the diffusion step at the required rate.

Similar calculations for the other data points enable the growth rate to be plotted against the driving force at the surface as in Figure 3.3.6, so enabling the integration kinetics to be evaluated. In this case the best power law equation (of the form of Equation (2.4.3)) is:

$$R_G = 3.93 \cdot 10^{-3} \cdot (w_I - w^*)^{1.51}$$

where R_G is in $\mathrm{kg\,m^{-2}\,s^{-1}}$ and w in kg sucrose per kg solution, and this relation is included in Figure 3.3.6.

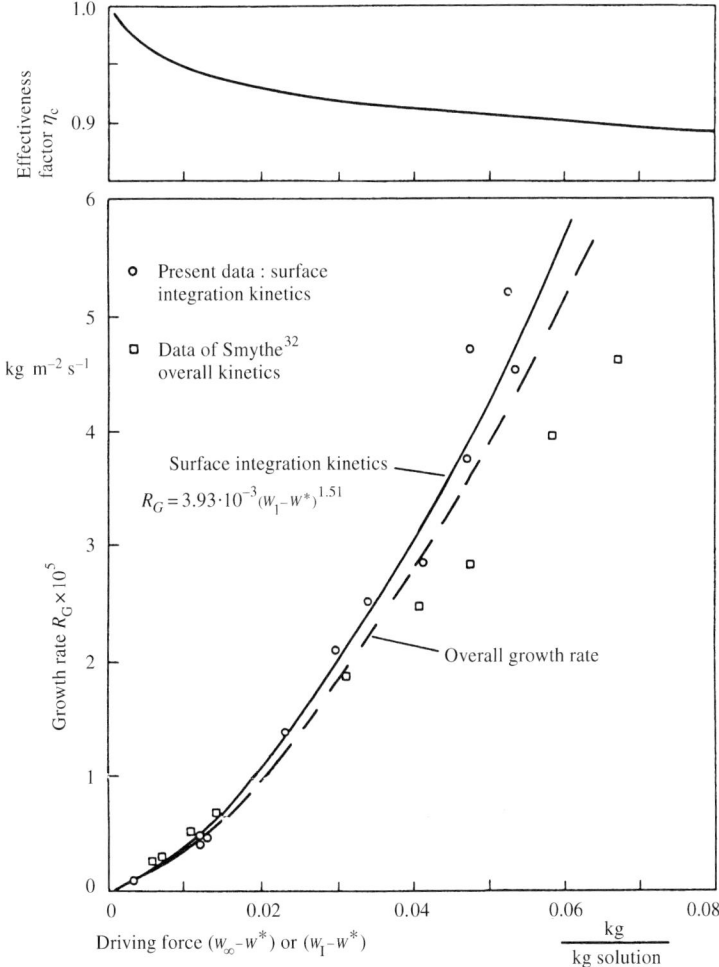

Figure 3.3.6 Surface integration and overall growth kinetics for sucrose at 31°C

When the overall rate is included on the same co-ordinates it is again clear that in this case the overall process is dominated by the integration kinetics. This is reinforced since the value of the effectiveness factors (Section 2.4.2) are between 0.97 and 0.90.

3.3.6 Comparison with other data

Growth kinetics were also measured by Smythe[32] for the system described above. These were obtained by measuring the increase in weight of a single crystal held in the discharge stream of an impeller situated in a small tank. By working at increasing impeller rotation speeds, Smythe claimed to determine the point at which growth rate was no longer influenced by diffusion effects.

Smythe's data for such a high rotation speed, also at 31.0°C, is included in Figure 3.3.6. Data points are very close to those from the rotating disc. Any inaccuracy arising from the difficulty of eliminating diffusion effects by this method would be masked in this case since the overall rate is dominated by the integration step in all but the most gently agitated solution.

3.4 Batch multiparticle systems

The main advantages of batch multiparticle experiments are:

- easy operation of batch experiments;
- use of a large amount of seeds leading to an average value of crystal growth rate suitable for design purposes. These values are independent of individual crystal properties such as slightly different shapes or different numbers of dislocations.

According to the size distribution of the seeds such systems can be subdivided into:

- monodisperse (or nearly monodisperse) systems (with a narrow CSD); and
- polydisperse systems (with a broad CSD).

According to the hydrodynamics they can be further classified as:

- fluidized bed methods; and
- agitated suspension methods;

while the temperature regime can be distinguished between:

- isothermal or quasi-isothermal methods; and
- polythermal methods.

With respect to the method of measurement the methods are subdivided into:

- pseudo-differential methods; and
- integral methods;

while, finally, according to the degree of access to the system, there are:

- direct methods through measurement of the increase of crystal mass or crystal size; and
- indirect methods from measurements using solution side information.

All possible combinations of these methods may be envisaged, but only a limited number of them are of practical value[64].

3.4.1 Fluidized bed measurement (weighing method)

The use of fluidized single crystals has the advantage that the crystals can rotate and move freely in the flowing solution making this method more representative of conditions found in real crystallizing processes. On the other hand, the relative velocity is always dependent on the particle size whereas for a fixed single crystal that cannot move or even rotate in the solution, relative velocity can be varied independently of crystal size.

The measurement using a larger number of crystals gives a certain average of individual crystal growth rates. Therefore the fluidized bed method gives the overall crystal growth rate and is not so good for checking growth theories. This limitation holds for all methods using multiparticle systems and giving overall growth rates.

(a) Principle of the method

A known amount N_{tot} of selected crystals sorted from a narrow sieve fraction and having initial weight $M_{C\alpha}$ is grown in an upflowing stream of solution with supersaturation equal to:

$$\Delta W = \Delta\vartheta \cdot \frac{dW^*}{d\vartheta} \tag{3.4.1}$$

for a time period t. The crystals are then separated and their final mass $M_{C\omega}$ is determined. The growth rate is:

$$G = \frac{M_{C\omega}^{1/3} - M_{C\alpha}^{1/3}}{(\alpha \cdot \rho_C \cdot N_{tot})^{1/3} \cdot t} \tag{3.4.2}$$

where α is the volume shape factor and ρ_C is the density of the crystals. Simultaneously, the following kinetic reaction holds:

$$G = \frac{\beta \cdot k_G''}{3 \cdot \alpha \cdot \rho_C} \cdot \Delta W^g \tag{3.4.3}$$

i.e.

$$\log G = \log\left(\frac{\beta \cdot k_G''}{3 \cdot \alpha \cdot \rho_C}\right) + g \cdot \log \Delta W \tag{3.4.4}$$

where k_G'' and g are the growth rate constant and the growth rate order, respectively. The value of g is found from the slope of a linear plot of $\log G$ versus $\log \Delta W$ (Equation 3.4.3) and k_G'' is then given by:

$$k_G'' = \frac{3 \cdot \alpha \cdot \rho_C \cdot G_i}{\beta \cdot \Delta W_i^g} \tag{3.4.5}$$

where G_i and ΔW_i are co-ordinates of an arbitrary point on the line.

(b) Equipment and measurement procedure

The schematic outline of the apparatus for fluidized bed measurements is depicted in Figure 3.4.1 (see page 81). A stock solution vessel (1) of 5 l capacity contains the solution maintained 1 K above its saturation (equilibrium) temperature, ϑ^*. It is thermostated using water from a commercial thermostat, circulated through a glass spiral cooler in vessel 1 (not depicted), or using an infralamp operated by a contact thermometer. Good mixing of the solution, together with its transfer, is executed by a centrifugal pump (2). The solution passes through a small thermometric vessel (3) fitted with a thermometer (4), where a recirculation tube is also connected into a cooler made of glass (5). The cooler is fed by water from another thermostat, so that the solution is cooled to ϑ_2 giving the required undercooling:

$$\Delta\vartheta = \vartheta^* - \vartheta_2 \tag{3.4.6}$$

This cooled solution then passes through a valve into the measuring tube (8) 500 mm long, with diameters of 50 mm in the upper calming section, 26 mm in the fluidized bed section and only 10 mm in the inlet neck. The temperature in the fluidized bed is measured using another thermometer.

Introduction and removal of seed crystals is performed using two pistons whose location and construction can be seen from Figures 3.4.1, 2 and 3 (see pages 81 and 82). Both pistons are made of Teflon and may be moved in a glass or perspex sleeve by pushing and turning the piston. They are secured in the tube by turned ribs and possess a recess for accommodating crystals. Introduc-

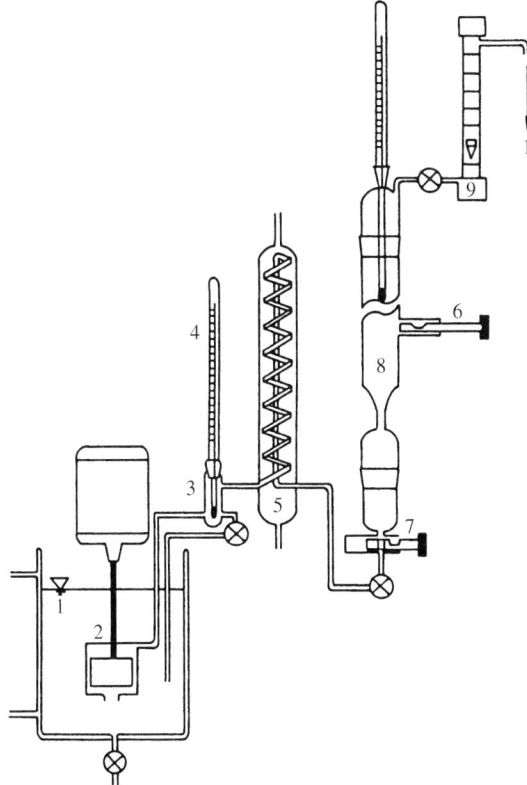

1. storage tank
2. pump
3. measuring tube
4. thermometer
5. cooler
6. input of seeds
7. output of crystals
8. measuring tube
9. flow meter

Figure 3.4.1   Apparatus for fluidized bed measurements

tion and removal of crystals can also be achieved without the pistons but the operation is more difficult and less accurate.

The seed crystals are initially classified by sieving and then selected under a microscope from a narrow sieve fraction. Only crystals of similar size and shape, having near-perfect faces are chosen. Each run starts with 20 to 60 seeds, briefly cured in a just-saturated solution, wiped on filter paper and weighed ($M_{C\alpha}$). The shape factors are determined simultaneously (see Section 2.3). The solution of measured undercooling is circulated, crystals are introduced into the recess of the upper piston while in the drawn position, the piston is pushed in and the crystals are thus washed out in the sheathed and turned position of the piston. After elapse of the desired measurement time, the discharge piston is turned through 90° so that the flow is reduced and the crystals pass down through the neck into the recess of the piston. After withdrawing the piston and rotating through 180° they may be removed. On replacing the piston into the

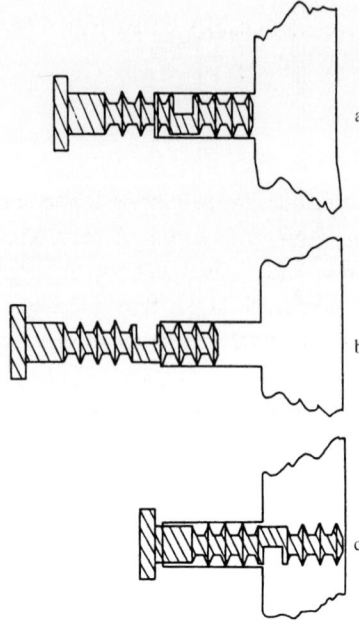

a. measurement position
b. feeding a crystal
c. introduction of seeds

Figure 3.4.2   Piston for introduction of seeds

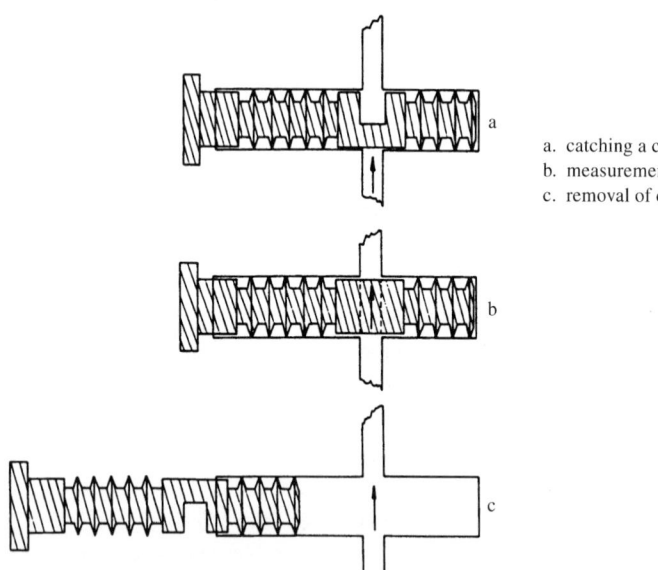

a. catching a crystal
b. measurement position
c. removal of crystals

Figure 3.4.3   Piston for the removal of crystals

original position, the flow of the solution enables a new experiment to start. Removed crystals are wiped and weighed ($M_{C\omega}$).

(c) Example

Growth rates of $CuSO_4 \cdot 5\ H_2O$ crystals were measured[65]. The seed crystals were taken from a sieve fraction $0.60-0.75$ mm and their shape factors determined as $\alpha = 0.288$ and $\beta = 3.01$. The crystals density is $\rho_C = 2286\ kg\ m^{-3}$. Data obtained for a superficial liquor flow velocity (calculated from the measured value of volumetric flow rate) in the measuring tube equal to 5 cm per second are presented in Table 3.4.1 on page 84 and depicted in Figure 3.4.4 (see page 85). The method of least squares applied to the data gives:

$$G = (11.1 \pm 0.7) \cdot 10^{-6} \cdot \Delta W$$

and

$$k_G'' = (7.3 \pm 0.5) \cdot 10^{-3}\ kg\ m^{-2}\ s^{-1}$$

(d) Example of individual calculations

Run 11:

Equation (3.4.1): $\quad \Delta W = (31.50 - 30.65) \cdot 8.94 \cdot 10^{-3}$
$$= 7.60 \cdot 10^{-3}\ kg\ per\ kg\ H_2O$$

Equation (3.4.2): $\quad G = \dfrac{(9.188 \cdot 10^{-5})^{1/3} - (4.045 \cdot 10^{-5})^{1/3}}{(0.288 \cdot 2286 \cdot 60)^{1/3} \cdot 3600}$
$$= 8.81 \cdot 10^{-8}\ m\,s^{-1}$$

Equation (3.4.4): $\quad -7.05502 = const. + g \cdot 2.11919$

Run 3:

Equation (3.4.1): $\quad \Delta W = (29.55 - 29.20) \cdot 8.69 \cdot 10^{-3}$
$$= 3.04 \cdot 10^{-3}\ kg\ per\ kg\ H_2O$$

Equation (3.4.2): $\quad G = \dfrac{(2.534 \cdot 10^{-5})^{1/3} - (2.130 \cdot 10^{-5})^{1/3}}{(0.288 \cdot 2286 \cdot 30)^{1/3} \cdot 1800}$
$$= 3.40 \cdot 10^{-8}\ m\,s^{-1}$$

Equation (3.4.4) $\quad -7.46852 = const. + g \cdot 2.51713$

Subtracting the two expressions derived from Equation (3.4.4) we obtain:

$$0.41350 = g \cdot 0.39794$$

Table 3.4.1 Data obtained for a superficial liquor flow velocity in the measuring tube equal to 5 cm s^{-1}

Run	ϑ^* °C	ϑ_2 °C	ΔW $*10^{-3}$ kg/kg H_2O	N_{tot}	t s	$M_{C\alpha}$ $*10^{-5}$ kg	$M_{C\omega}$ $*10^{-5}$ kg	G $*10^{-8}$ m s^{-1}
1	30.25	30.10	1.33	30	1800	2.723	2.984	1.92
2	30.25	29.90	3.10	30	1800	2.868	3.378	3.53
3	29.55	29.20	3.04	30	1800	2.130	2.534	3.40
4	29.55	28.55	8.70	30	1800	3.195	4.635	8.61
5	29.55	29.10	3.91	30	1800	2.182	2.682	4.09
6	29.55	28.55	8.70	30	1800	2.410	3.725	9.27
7	29.40	28.90	4.34	60	3600	3.898	6.660	5.41
8	29.90	28.40	13.19	60	3600	3.845	11.910	12.60
9	29.90	28.40	13.19	60	3600	3.895	13.040	13.70
10	29.90	28.45	12.75	60	3600	4.060	14.740	15.00
11	31.50	30.65	7.60	60	3600	4.045	9.188	8.81
12	31.50	30.60	8.32	60	3600	3.965	10.770	10.47
13	31.50	30.55	8.70	60	3600	3.990	10.082	10.09
14	31.50	30.50	9.00	60	3600	3.905	11.070	11.49
15	31.50	30.45	9.35	60	3600	3.860	10.450	10.85
16	31.50	30.40	9.92	60	3600	3.857	9.500	9.66

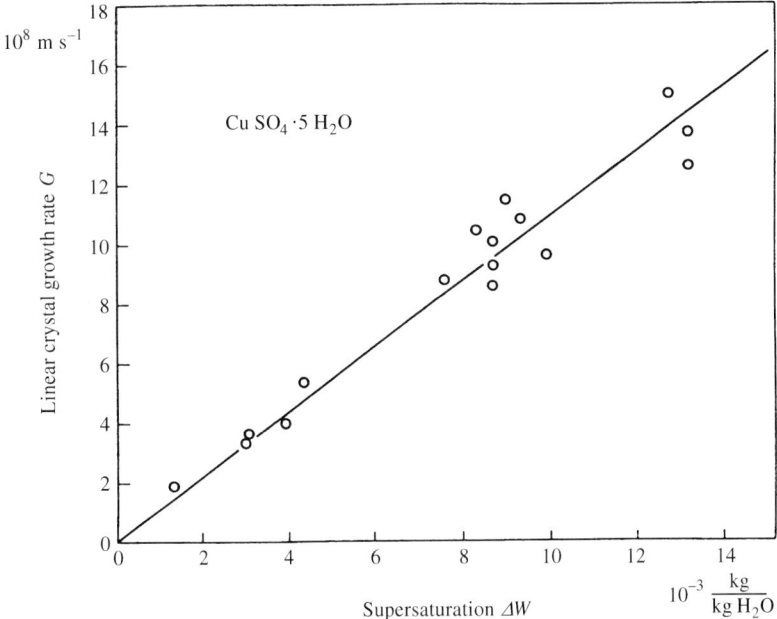

Figure 3.4.4   Growth rate of $CuSO_4 \cdot 5\,H_2O$ (example)

hence:

   $g = 1.04$

(simultaneous evaluation of all measurements gives $g = 1.00$). From Equation (3.4.5) for run 11:

$$k_G'' = \frac{3 \cdot 0.288 \cdot 2286 \cdot 8.81 \cdot 10^{-8}}{3.01 \cdot 7.60 \cdot 10^{-3}} = 7.61 \cdot 10^{-3}\,\mathrm{kg\,m^{-2}\,s^{-1}}$$

Evaluation of the data from all the experiments leads to:

   $G = 10.9 \cdot 10^{-6} \cdot \Delta W^{1.0}$

   $k_G'' = 7.2 \cdot 10^{-3}\,\mathrm{kg\,m^{-2}\,s^{-1}}$

3.4.2 Fluidized single crystals (video camera)

(a) Principle

The fluidized single crystals are suspended in the solution using a small conical transparent channel. Here they can be observed by microscope which may be combined with a video installation. The progress of the growing crystal is recorded on the video together with time data; thus it is possible to measure the elongation or shortening (the latter in the case of dissolution experiments) of specific crystal dimensions, stopping the recorder just when the required crystal face has a position parallel to the screen surface.

Usually the experiments are carried out at isobaric and isothermal conditions.

(b) Equipment and measurement procedure

The equipment consists of three main parts (see Figure 3.4.5):

- The experimental circuit comprising a solution reservoir, double wall tube heat exchangers, a circulation pump and thermometers for the exact control of temperature, and thus of supersaturation, in the fluidizing zone.
- The fluidizing zone — a small transparent conical channel (growth cell) where the fluidized crystal(s) can be observed.
- An inverse microscope combined with a TV camera, an adequate video recorder and a high resolution screen.

A number of details of the equipment are important:

- Temperatures are measured by Pt 100 — probes producing a reading accuracy of the installation of ±0.01 K. The probe diameter is 1.9 mm.

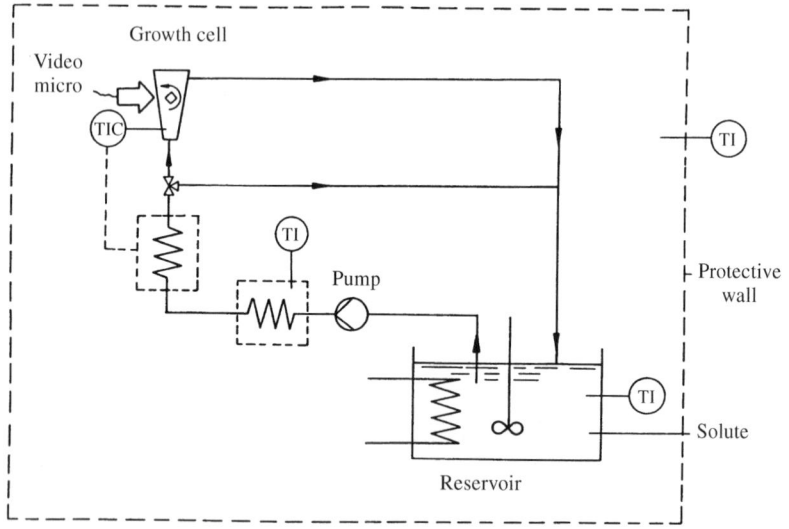

Figure 3.4.5 Schematic diagram of fluidized single crystal experimental arrangement

- The solution is circulated by a small centrifugal pump. To avoid sealing problems it is advantageous to use pumps with a magnetic clutch. Typical specifications are a driving power of about 50 W producing a flow rate of about 40 l per min.
- Thermostats in the range of 0−2000 W and 20−100°C are required and a constancy of temperature of 0.01 K at 70°C is needed.
- To avoid corrosion, all elements which are in contact with the solution should be made from glass, PMMA (Polymethylmethacrylate), PTFE (Polytetrafluorethylene) respectively coated with PTFE. Only the temperature probes are made from stainless steel, as they are not available in the above mentioned materials.
- The growth cell (see Figure 3.4.6), made from PMMA, is constructed in the lower part as a convergent channel.

Typical geometric dimensions are:
inlet opening:	2 mm × 32 mm
outlet opening:	8 mm × 32 mm
height of the convergent part:	35 mm
height of the upper part, with the same sectional area as the outlet opening:	100 mm

The growth cell is closed by very fine screens (60 μm) at both ends so that the crystals cannot move away from the cell as long as they are larger than the

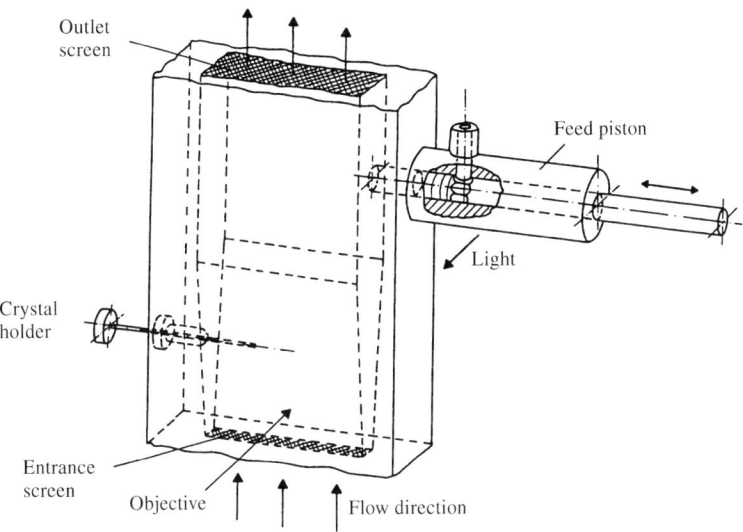

Figure 3.4.6 Growth cell observation of fluidized single crystals

mesh size. A small feed piston is located at one side of the cell to load the seed(s) into the cell.

A needle (crystal holder) can be installed near the position where the fluidized crystal is expected. After fixing a crystal at its tip, the needle is brought into position. Before starting the actual growth experiment, this crystal is observed through the microscope to determine the temperature at which the crystal neither grows nor dissolves; defining the saturation temperature. The temperature of the solution is then reduced at the cell entrance to generate the required supersaturation.

The region of the cell where the crystal is suspended is illuminated by a cold light source to avoid any thermal disturbances and to verify optimal conditions for the video microscope. Usually crystals under observation are in the size range of $250-1000 \, \mu m$, but predominantly $400-500 \, \mu m$.

The video equipment consists of the following elements (Figure 3.4.7):

- A video camera which should be able to present a high resolution.
- A video recorder, which should enable the user to stop the tape step by step in order to find the best position for the crystal to be observed and thus to measure its linear extension.
- A timer, which superimposes digital indications of time and data on the picture in the television system.

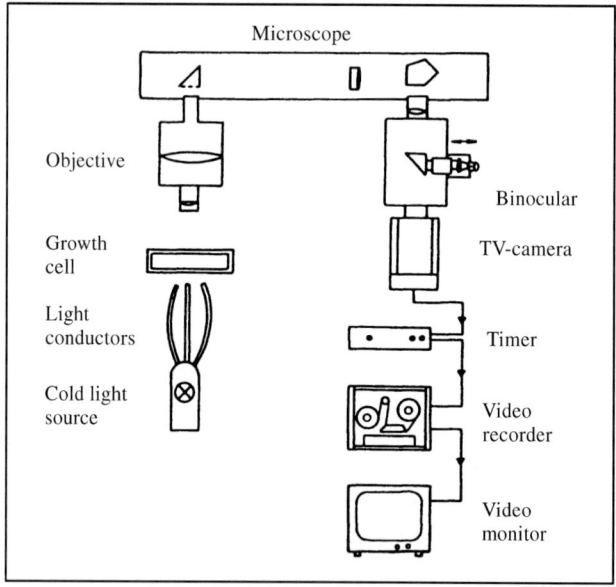

Figure 3.4.7 Video-microscope installation

The suspended crystal is observed by an inverse microscope. As there is always a distance of at least 15 mm between the crystal and the objective aperture and as it is quite difficult to focus the crystal when starting the observations, the camera should be equipped with a zoom-objective (1.8/20–100 mm).

The solution of the required concentration is prepared at some degrees above its saturation temperature, filtered and then charged into the solution reservoir. The saturation point is fixed by lowering the temperature or adding very small quantities of water or salt to the reservoir.

To reach the required supersaturation the temperature is lowered by cooling the solution using the heat exchangers located between pump and growth cell. The seed crystal is loaded into the cell by the feed piston, focused with the microscope and observations recorded online. It is recommended that the linear crystal extension versus time is measured from the recorded data (and online) after the real experiment is finished.

During this process the amount of the crystallized mass is extremely low compared with the whole solute mass in the system. Therefore, one may assume that steady state conditions are realized for a sufficient large period of time.

(c) Example

The example described here refers to the growth of fluidized single sodium chloride crystals from aqueous solutions.

The solubility curve of sodium chloride in water can be described by the following equation[66,67]:

$$W^*(\vartheta) = \frac{35.549 - 0.23125 \cdot \vartheta}{1 - 0.0069163 \cdot \vartheta} \quad \text{in the range } 30\text{--}80^\circ C$$

with:

$$\vartheta \text{ in } [^\circ C] \quad \text{and} \quad W^*(\vartheta) \text{ in } \left[\frac{\text{g NaCl}}{100 \text{ g water}} \right]$$

A part of this curve is shown in Figure 3.4.8 (see page 90).

From this figure it can be seen that a reduction of about $5^\circ C$ in the temperature of a NaCl solution, which is saturated at $55.13^\circ C$, generates a supersaturation of $\Delta W = 0.185$ g NaCl/100 g H_2O. On the screen of the TV monitor the increase of the length of cubic crystals was measured to be $\Delta L = 0.204$ mm. As the growth period in this case was 54 min, a linear growth rate of:

$$G = 6.3 \cdot 10^{-8} \, \text{m s}^{-1}$$

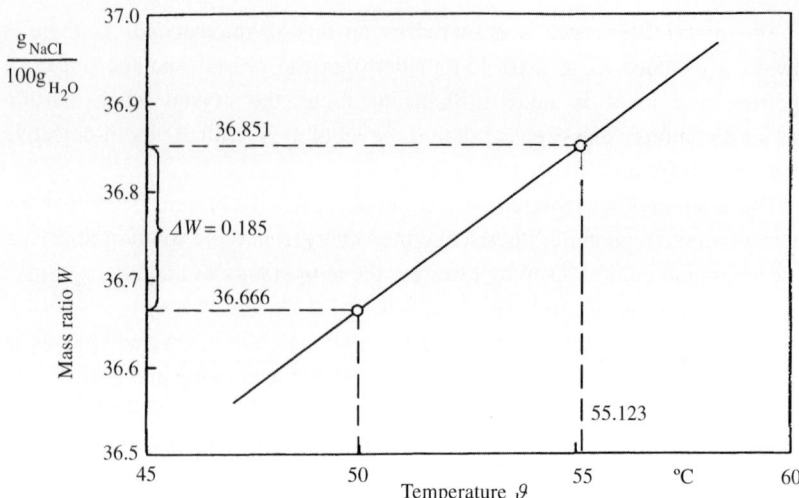

Figure 3.4.8 Detail of the NaCl–H_2O solubility-curve

is obtained. Since cubic crystals were observed, it can be assumed that the shape factors α is 1.0 and β is 6.0 (see Chapter 2.6). Thus with Equation (2.3.4) the mass flow density R_G can be calculated:

$$R_G = \frac{1}{2} \cdot \rho_C \cdot G = 0.68 \cdot 10^{-4}\,\mathrm{kg\,m^{-2}\,s^{-1}}$$

3.4.3 Batch differential growth

This method has some similar features to the fluidized bed measurements described above although the measurement is carried out in an agitated vessel. The apparatus is thus much simpler but the accuracy of the measurement is somewhat less.

(a) Principle of the method

A large amount of solution of known saturation temperature, ϑ^*, is cooled to temperature ϑ_2 within the metastable zone so that the supersaturation is given by

$$\Delta W = (\vartheta^* - \vartheta_2) \cdot \frac{dW^*}{d\vartheta} \tag{3.4.7}$$

A small amount of selected seed crystals N_{tot} is added to the solution for a time period t. The crystals are then removed and their mass or size increase is determined. The mass increase must be small enough for the supersaturation to be considered constant.

The basic kinetic equation for the mass increase of crystals is:

$$M_C = k''_G \cdot A_C \cdot \Delta W^g \tag{3.4.8}$$

where A_C is the surface area of N_{tot} crystals, therefore

$$A_C = \beta \cdot N_{tot} \cdot L^2 \tag{3.4.9}$$

For a small size increase we may consider A_C to be constant and thus

$$\frac{M_{C\omega} - M_{C\alpha}}{t} = k''_G \cdot \beta \cdot N_{tot} \cdot L^2 \cdot \Delta W^g \tag{3.4.10}$$

If the size increase cannot be neglected, we can take as an approximation[68]:

$$\bar{A}_C = \frac{1}{L_\omega - L_\alpha} \int_{L_\alpha}^{L_\omega} A_C \cdot \frac{L}{L_\alpha} \cdot dL = \frac{1}{3} \cdot A_{c\alpha} \left[1 + \left(\frac{M_{C\omega}}{M_{C\alpha}} \right)^{1/3} + \left(\frac{M_{C\omega}}{M_{C\alpha}} \right)^{2/3} \right] \tag{3.4.11}$$

so that

$$\frac{M_{C\omega} - M_{C\alpha}}{t} = k''_G \cdot \beta \cdot N_{tot} \cdot L_\alpha^2 \cdot \left[1 + \left(\frac{M_{C\omega}}{M_{C\alpha}} \right)^{1/3} + \left(\frac{M_{C\omega}}{M_{C\alpha}} \right)^{2/3} \right] \cdot \Delta W^g \tag{3.4.12}$$

(b) Equipment and measurement procedure

The measurements are best performed in a three-necked round bottom flask of capacity between 2 and 4 l, equipped with a stirrer and a thermometer. The flask is placed in a thermostated vessel whose temperature is controlled with an accuracy better than ± 0.05 K. After stabilizing the temperature, selected seed crystals (N_{tot}, $M_{C\alpha}$) are introduced into the solution of known supersaturation (Equation 3.4.1). After a batch time t has elapsed, crystals are removed, dried and reweighed ($M_{C\omega}$). The hydrodynamics of the system are characterized by, for example, the Reynolds number of the stirrer

$$Re_{stirrer} = \frac{s \cdot d^2 \cdot \rho_L}{\eta_L} \tag{3.4.13}$$

or the Reynolds number of the crystals (particles) with size L

$$Re_p = \frac{L^{4/3} \cdot (P/M_{sus})^{1/3} \cdot \rho_L}{\eta_L} = \frac{L^{4/3} \cdot (\bar{\varepsilon})^{1/3} \cdot \rho_L}{\eta_L} \tag{3.4.14}$$

where s is the rotational speed (s^{-1}), d is the diameter of the stirrer (m), ρ_L and η_L are the density (kg m^{-3}) and viscosity (Pa s) of the solution, respectively and $\bar{\varepsilon} = P/M_{sus}$ the energy dissipated by the stirrer per unit mass of suspension ($W\,kg^{-1}$). The growth order g and the growth rate constant k_G'' are calculated from a plot according to the equation:

$$\log\left(\frac{M_{C\omega} - M_{C\alpha}}{t}\right) = const. + g \cdot \log \Delta W \tag{3.4.15}$$

and Equations (3.4.10) or (3.4.12), respectively.

(c) Example

A 4 l flask was filled with a solution of copper sulphate saturated at $\vartheta^* = 32.5°C$ and thermostated at $31.0°C$. From the sieve fraction $1.00-1.15\,mm$ 50 crystals of $CuSO_4 \cdot 5\,H_2O$ with well developed faces were selected. Their surface area shape factor is 3.01 and their total weight $M_{C\alpha} = 3.875 \cdot 10^{-5}\,kg$. These seeds were introduced into the supersaturated solution and the vessel was stirred by a two-blade agitator ($d = 4\,cm$, $P_o \approx 0.9$) at a rotation speed of $6.67\,s^{-1}$. The density of the solution is $1220\,kg\,m^{-3}$ and viscosity $2.5 \cdot 10^{-3}\,Pa\,s$. After 15 mins the crystals were filtered off, dried and weighed ($M_{C\omega} = 5.445 \cdot 10^{-5}\,kg$). From previous experiments it is known that $g = 1$.

Equation (3.4.10) then gives:

$$k_G'' = \frac{5.445 \cdot 10^{-5} - 3.875 \cdot 10^{-5}}{900 \cdot 3.01 \cdot 50 \cdot 1 \cdot 10^{-6} \cdot (32.5 - 31.0) \cdot 9.07 \cdot 10^{-3}}$$
$$= 8.52 \cdot 10^{-3}\,kg\,m^{-2}\,s^{-1}$$

This value has been found for agitation characterized by the stirrer Reynolds number according to Equation (3.4.13):

$$Re_{stirrer} = \frac{6.67 \cdot (0.04)^2 \cdot 1220}{2.5 \cdot 10^{-3}} = 5208$$

and the particle Reynolds number according to Equation (3.4.14) with

$$\bar{\varepsilon} = \frac{4 \cdot P_o}{\pi} \cdot s^3 \cdot d^2 \cdot \left(\frac{d}{D}\right)^2 \cdot \left(\frac{d}{H}\right) = \frac{Ne \cdot s^3 \cdot d^5}{V} = 0.007\,W\,kg^{-1}$$

$$Re_p = 10.2$$

3.4.4 Desupersaturation of solution

The rate of mass transfer from a liquid to the solid phase can also be measured from information obtained from the solution. This method has the advantages that the measurements can be carried out under conditions similar to those existing in a crystallizer and growth data as a function of supersaturation can be obtained from one experiment. The main disadvantages are the necessity of knowing the crystal surface area and of making a series of solution composition analyses[69].

(a) Principle of the method

A large number of seed crystals of known surface area are added to the solution of initial supersaturation ΔW_α. The crystals start to grow immediately and hence reduce the supersaturation of the solution. The supersaturation can be monitored continuously using a physical method (solution density, conductivity, refractive index for example), or small samples of the solution can be taken at short time intervals and analysed by a suitable physical or chemical method.

Assuming that no nucleation occurs, the change in the supersaturation in a closed isothermal system is given by the mass balance equation:

$$\frac{d(\Delta W)}{dt} = -k''_G \cdot A_C \cdot \Delta W^g \tag{3.4.16}$$

If the overall crystal surface area does not change substantially during the experiment, Equation (3.4.16) can be integrated to give:

$$k''_G = \frac{1}{(g-1) \cdot A_C \cdot t} (\Delta W_t^{1-g} - \Delta W_\alpha^{1-g}) \tag{3.4.17}$$

or, for $g = 1$:

$$k''_G = \frac{1}{A_C \cdot t} \cdot \ln\left(\frac{\Delta W_\alpha}{\Delta W_t}\right) \tag{3.4.18}$$

where ΔW_t is the supersaturation at time t.

The surface area of crystals can be determined from the size of the seed crystals and their weight:

$$A_C = \frac{\beta \cdot M_C}{\alpha \cdot \rho_C \cdot L} \tag{3.4.19}$$

or by using Equation (3.4.11). This approximation is the main source of error in this method.

(b) Equipment and measurement procedure

A $4\,dm^3$ capacity round-bottommed flask, equipped with a stirrer, an accurate thermometer and a draft tube is submerged in a water bath whose temperature is operated with an accuracy of $\pm 0.05°C$. A solution of the substance to be investigated is just saturated at temperature ϑ^* and then cooled to temperature ϑ_2 in order to obtain the desired initial supersaturation ΔW_α:

$$\Delta W_\alpha = (\vartheta^* - \vartheta_2) \cdot \frac{dW^*}{d\vartheta} \qquad (3.4.20)$$

At time $t = 0$, a known amount of seeds $(M_{C\alpha}, L_\alpha)$ is introduced and measurement of supersaturation is started immediately, either continuously or by sampling at 1 minute intervals.

The growth order can be obtained from a logarithmic plot according to:

$$\log\left(\frac{d(\Delta W)}{dt}\right) = \text{const.} + g \cdot \log \Delta W \qquad (3.4.21)$$

and hence from the slope of the line obtained by plotting $\log(-d(\Delta W)/dt)$ against $\log (\Delta W)$. The left-hand side can be most easily found if the sampling was in equal time intervals. The kinetic growth constant k_G'' is then calculated using Equations (3.4.17) or (3.4.18), respectively.

(c) Example

Ammonium sulphate solution[69] saturated at 60°C was placed into a 4 dm³ crystallizer and cooled to 56°C. 75 g of crystals from a narrow sieve fraction of 0.5 to 0.6 mm were placed in the solution at time $t = 0$. Samples were taken at 1 minute intervals using a 10 cm³ pipette (preheated in a drying oven in order to prevent crystallization) with a removable cotton filter, poured into series of weighing bottles and weighed $(M_1)_t$. The opened weighing bottles were then placed in a drying oven at 80°C and the water completely evaporated, the temperature of the drying oven raised to 120°C and the final (constant) weight of samples $(M_2)_t$ determined after 10 hours. The final solution sample was taken after 20 minutes to check that it corresponded to the equilibrium value finally reached (subscript ω). The supersaturation of individual samples is then:

$$\Delta W_t = \left(\frac{M_2}{M_1 - M_2}\right)_t - \left(\frac{M_2}{M_1 - M_2}\right)_\omega \qquad (3.4.22)$$

Experimental data are summarized in Table 3.4.2 (see page 95).

Calculation of A_C: 4 l of solution multiplied by the solution density $\rho_L = 1247 \text{ kg m}^{-3}$ corresponds to 5.0 kg of solution and contains $M_{H_2O} = 5/(1 + W_\alpha) = 5/1.8721 = 2.67$ kg water. The mass of seeds is $M_{C\alpha} = 75 \cdot 10^{-3}/2.67 = 28.09 \cdot 10^{-3}$ kg per kg H_2O. Hence:

$$A_C = \frac{6 \cdot 28.09 \cdot 10^{-3}}{1 \cdot 1769 \cdot 5.5 \cdot 10^{-4}} = 0.1732 \text{ m}^2 \text{per kg } H_2O$$

Table 3.4.2 Summary of experimental data

Sample	t	M_1	M_2	ΔW	$d\Delta W/dt$	k_G''
	s	g	g	$*10^{-4}\,kg/kg$	$*10^{-5}\,kg/kg \cdot s$	$g=1.66$
1	0	12.4815	5.8143	133	−8.0	
2	60	12.4532	5.7841	85	−4.8	0.89
3	120	12.5106	5.8006	56	−2.83	0.99
4	180	12.4251	5.7546	39	−2.5	1.07
5	240	12.4805	5.7749	24	−0.83	1.35
6	300	12.4857	5.7755	19	−0.21	1.35
7	1200	12.4700	5.7614	0	−	−

k_G'' (mean) $= 1.13$

Calculation of the supersaturation is shown for data of sample No. 4:

Equation (3.4.22):
$$\Delta W = \frac{5.7546}{12.4251 - 5.7546} - \frac{5.7614}{12.4700 - 5.7614}$$
$$= 0.8627 - 0.8588 = 0.0039 \text{ kg per kg } H_2O$$

Calculation of $d(\Delta W)/dt$ for sample No. 4:

$$\frac{d(\Delta W)}{dt} = \frac{39 \cdot 10^{-4} - 24 \cdot 10^{-4}}{180 - 240} = -2.5 \cdot 10^{-5} \text{ kg per kg } H_2O$$

Calculation of the kinetic exponent g, for sample 1 through 5 from Equation (3.4.21) using the method of least squares, leads to the equation:

$$\log\left(-\frac{d(\Delta W)}{dt}\right) = -0.85 + 1.6645 \cdot \log(\Delta W)$$

and hence g $= 1.66$ and Equation (3.5.17) can be used for calculation of k_G''.

3.4.5 Quasi-isothermal model crystallization

This method has the advantage that it is possible to determine the growth rate as a function of supersaturation in one simple experiment, provided that either the supersaturation or the mass deposition during the experiment can be measured. One of the possibilities of monitoring the mass deposition is by calorimetry[70-72].

(a) Principle of the method

At the beginning of the experiment the calorimeter contains a known volume of supersaturated solution and a known mass of crystals in a sealed ampoule. The temperature variation in the calorimeter is divided into the initial (before the process of crystallization is initiated), main (after breaking the ampoule) and final periods (Figure 3.4.9). In the initial period, slow temperature changes in the system occur due to the energy dissipated by the stirrer and to the rate of heat exchange between the calorimeter and its jacket. This initial period can be described by the equation:

$$C \cdot \frac{d\vartheta}{dt} = \dot{Q} \tag{3.4.23}$$

where the heat capacity of the calorimetric system C can be determined by an independent calibration using electric resistance heating as the source of Joule heating:

$$C \cdot \Delta\vartheta = \dot{Q} \cdot t + E \cdot I \cdot t \tag{3.4.24}$$

where E and I are the voltage and current intensity applied for heating during the period t. $\dot{Q} \cdot t$ can usually be neglected.

The crystallization process is initiated by breaking the ampoule containing crystals at time $t = t_\alpha$. The heat balance must now include the thermal effect of crystallization:

$$C \cdot \frac{d\vartheta}{dt} = -\frac{\rho_C \cdot \Delta H_C}{\tilde{M}_{hyd}} \cdot G \cdot A_C + \dot{Q} \tag{3.4.25}$$

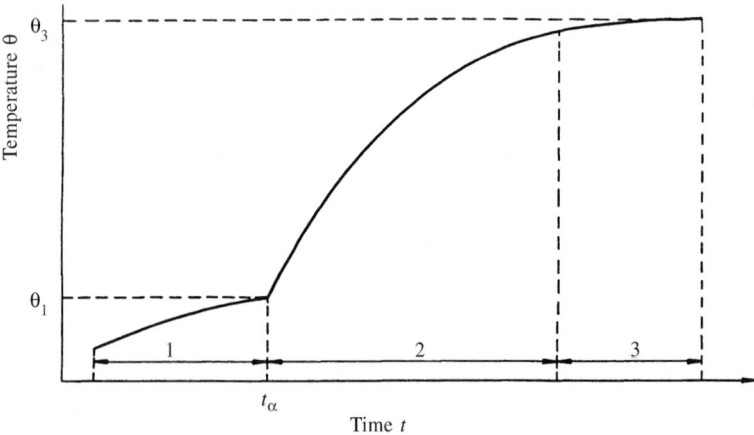

Figure 3.4.9   Schematic draft of a calorimetric curve: 1 — initial period, 2 — main period, 3 — final period

where ρ_C is the crystal density, A_C the total crystal surface area and ΔH_C the heat of crystallization (negative for most salts). The temperature variation in the final period is determined by the equation

$$\left(C - \frac{\Delta H_C \cdot M_{H_2O}}{\tilde{M}_{hyd}} \cdot \frac{dW^*}{d\vartheta}\right) \cdot \frac{d\vartheta}{dt} = \dot{Q} \tag{3.4.26}$$

where M_{H_2O} is the mass of the free water in the system.

From the temperature variation in the initial and final periods it is possible to evaluate the characteristic temperatures ϑ_1 and ϑ_3. The instantaneous value of supersaturation at time t is:

$$\Delta W(t) = \frac{\tilde{M}_{hyd}}{\Delta H_C \cdot M_{H_2O}} \left[C \cdot (\vartheta - \vartheta_3) - \int_{t_3}^t \dot{Q} \cdot dt \right] - (\vartheta - \vartheta_3) \cdot \frac{dW^*}{d\vartheta} \tag{3.4.27}$$

while the mass deposited on the crystals is:

$$M_C = M_{C\alpha} + M_{H_2O} \cdot (W_\alpha - W) = \frac{1}{4} \cdot \alpha \cdot \rho_C \cdot q_o \cdot (L_{max}^4 - L_{min}^4) \tag{3.4.28}$$

and their surface area:

$$A_C = \frac{1}{3} \cdot \beta \cdot q_o \cdot (L_{max}^3 - L_{min}^3) = \frac{4 \cdot \beta \cdot M_C}{3 \cdot \alpha \cdot \rho_C} \cdot \frac{L_{max}^3 - L_{min}^3}{L_{max}^4 - L_{min}^4} \tag{3.4.29}$$

The growth rate can then be easily calculated from:

$$G = (A_C \cdot \rho_C)^{-1} \cdot \frac{dM_C}{dt} \tag{3.4.30}$$

In these equations it is assumed that the shape factors and the difference $(L_{max} - L_{min})$ are constant and that growth rate is size independent.

(b) Equipment and measurement procedure

Either a commercial calorimeter or a simple Dewar calorimetric vessel equipped with a stirrer and a semiconductor temperature sensor can be used for measurements. For example a twin calorimeter of the Dewar type fitted with a mechanical device for breaking a thin-walled glass ampoule containing crystals is used. A small amount of just-saturated solution is agitated with a vibrational stirrer at a frequency of 50 Hz. Thermistors in the bridge connection and the output for a line recorder enabled the temperature difference between two identical cells to be measured with a precision of ± 0.001 K.

The solution, saturated at 25°C, is heated to 80°C for 30 minutes, filtered and rapidly cooled to a temperature below 25°C. 220 cm^3 of the solution is placed in the calorimetric vessel. After starting the stirrer and reaching the stationary temperature response, which takes about 10 minutes, the initial temperature variation is recorded. This usually produces a rise of about $10^{-4}\,\mathrm{K\,s^{-1}}$. As the temperature rise during the initial period corresponds to the usual value of q, the main period is started by breaking the ampoule containing seed crystals. The end of the experiment is clear from the onset of the final temperature period. By using manganine resistance heating wire (130 ohm, 0.12 A) as the source of Joule heating, the thermal capacity of the system is found in such a way that the temperature rise in the Joule heating experiment is approximately equal to that in the crystallization experiment.

For evaluation of the temperature course, about 100 points are read from the recorded curves. The values $d\vartheta/dt$ are calculated by differentiating second-order polynomials fitted successively through each experimental point and a chosen number of its neighbours using the method of least squares. The supersaturations (Equation 3.4.27) and growth rates (Equations 3.4.28 through 2.4.30) are then calculated and plotted.

(c) Example
Crystals of $Na_2HPO_4 \cdot 12H_2O$ of size 0.25 to 0.4 mm were taken as seeds. 2.66 g of these were sealed into the ampoule with $1\,\mathrm{cm}^3$ of saturated solution. $220\,\mathrm{cm}^3$ of solution saturated at 25°C was cooled to $\vartheta_\alpha = 24.10°C$ and placed in the calorimeter. After 15 minutes the ampoule was broken and the experiment started. The duration of the main period was 375 seconds. Values of $\dot{Q}_1$ and $\dot{Q}_3$ were 0.282W and 0.260W respectively.

Data for evaluation:

$\Delta H_C = 94.20\,\mathrm{kJ}$ per mole, $\rho_C = 1520\,\mathrm{kg\,m^{-3}}$, $\beta/\alpha = 7.09$, $\rho_L = 1093\,\mathrm{kg\,m^{-3}}$, $\Delta W_\alpha = 4.4 \cdot 10^{-2}\,\mathrm{kg}$ per kg H_2O, the mean thermal capacity was $C = 1123\,\mathrm{J\,K^{-1}}$.

Results of the evaluation are shown in Figure 3.4.10 (see page 99) and correspond to the equation:

$$\log G = -2.72 + 1.27 \cdot \log \Delta W$$

A much simpler though not so exact procedure is as follows[72]:

The solution is placed into a jacketed vessel equipped with a stirrer, electrical heating and a thermistor thermometer. The vessel is filled with a known amount of solution saturated exactly at the measurement temperature. Mass M_C of crystals are added and dissolved at an elevated temperature. The measurement temperature is then established. At time t_α seeds are added and the temperature-time curve, ϑ (t) recorded. Now define:

$$\Delta\vartheta(t) = \vartheta(t) - \vartheta_j \qquad (3.4.31)$$

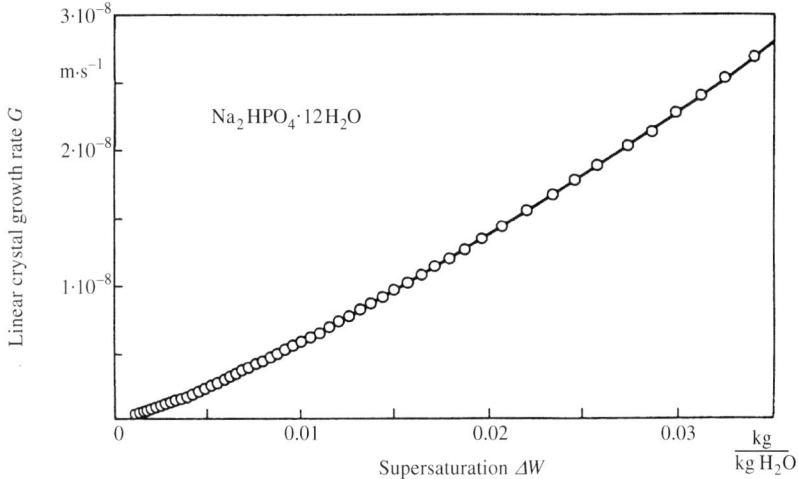

Figure 3.4.10   Growth rate of $Na_2HPO_4 \cdot 12H_2O$ (example)

where ϑ_j is the temperature of the surroundings, in this case the jacket. The heat flux:

$$\dot{Q}(t) = \frac{dQ}{dt}(t) = \frac{\varDelta\vartheta(t)}{R} + \frac{d(\varDelta\vartheta)}{dt}(t) \cdot C \qquad (3.4.32)$$

where R and C are system constants determined in a separate blank experiment. The values of $\dot{Q}(t)$ are then corrected for the heat produced by the stirrer and for the heat exchange with the surroundings:

$$Q = \int_{t_\alpha}^{t} \dot{Q}(t)dt \qquad (3.4.33)$$

From the total heat evolved during the crystallization and the final mass of crystals, the heat of crystallization:

$$\varDelta h_C = \frac{Q}{M_{C\omega}} \qquad (3.4.34)$$

can be obtained and, using this value, the mass of crystals deposited at every instant determined by the expression:

$$\dot{M}_C(t) = \frac{\dot{Q}(t)}{\varDelta h_C} \qquad (3.4.35)$$

99

or

$$M_C(t) = \frac{Q(t)}{\Delta h_C} \qquad (3.4.36)$$

Simultaneously the following equation holds:

$$M_C(t) = M_{H_2O} \cdot (\Delta W_\alpha - \Delta W(t)) \qquad (3.4.37)$$

So, knowing the initial supersaturation and the instantaneous solubility $W^*(t)$, the supersaturation is obtained. In this way, a plot of $\dot{M}_C$ versus ΔW can be constructed. The linear growth rate is then:

$$G = \frac{\dot{M}_C(t)}{3 \cdot M_C^{2/3}(t) \cdot (\alpha \cdot \rho_C \cdot N_{tot})^{1/3}} \qquad (3.4.38)$$

3.4.6 Polythermal model crystallization

Batch crystallization experiments have many advantages over continuous crystallization, particularly in that simple equipment and operation are employed. These advantages are, however, somewhat offset by difficulties in evaluating the experimental results and in a poorer reproducibility of measurements. One possibility of overcoming these difficulties is batch operation at constant supersaturation. The main disadvantage of the method is that only a mean growth rate over the whole temperature interval is obtained.

(a) Principle of the method

Consider a batch crystallizer of volume V provided with an efficient stirrer maintaining uniformity of temperature and concentration throughout the crystallizer and capable of maintaining the crystals in contact with the entire liquid volume. The crystallizer is filled with a solution saturated at temperature ϑ_α having mass:

$$M_\alpha = V \cdot \rho_L \qquad (3.4.39)$$

Suppose that this solution is cooled to temperature ϑ_ω in such a way that the supersaturation remains constant during the run. From the material balance on a cooling crystallizer it follows that for the solid to be precipitated:

$$\Delta M_C = M_\alpha \cdot \frac{W_\alpha - W_\omega}{W_\alpha + 1} \qquad (3.4.40)$$

The solution is cooled to form the anticipated supersaturation and seeded with a mass of crystals $M_{C\alpha}$ of size L_α sufficient to allow an acceptable rate of crystallization and, at the same time, to enable the crystals to grow to the required size L_ω. Thus:

$$M_{C\alpha} = \frac{L_\alpha^3}{L_\omega^3 - L_\alpha^3} \cdot \Delta M \qquad (3.4.41)$$

where L_ω = final size of seeds and ΔM = the mass increase resulting from the materials balance.

The seeded solution is now supersaturated at the maximum rate possible without significant nucleation occurring. The balance describing mass transfer between the liquid and solid phases can be written as:

$$-\left(\frac{d\vartheta}{dt}\right) \cdot \left(\frac{dW^*}{d\vartheta}\right) = k_G'' \cdot A_C \cdot \Delta W^g \qquad (3.4.42)$$

where the left-hand side describes the supersaturation rate and the term on the right-hand side describes the desupersaturation rate due to crystal growth. A_C is the crystal surface area and ΔW the supersaturation. Thus the allowable cooling rate $-\dot{\vartheta} = d\vartheta/dt$ is proportional to the surface area of crystals present at any instant[73,74]. The size of crystals at time t is:

$$L = L_\alpha + G \cdot t \qquad (3.4.43)$$

and, knowing their number:

$$N_{tot} = M_\alpha (\alpha \cdot \rho_C \cdot L_\alpha^3)^{-1} \qquad (3.4.44)$$

we can obtain their surface area at time t:

$$A_C = N_{tot} \cdot \beta \cdot L^2 = \frac{M_\alpha \cdot \beta}{\alpha \cdot \rho_C \cdot L_\alpha} \cdot \left(1 + \frac{G \cdot t}{L_\alpha}\right)^2 \qquad (3.4.45)$$

Substitution of this surface area into Equation (3.4.42) leads to the expression of the allowable cooling rate:

$$-\dot{\vartheta} = \frac{3}{M_{H_2O}} \cdot \frac{M_\alpha}{dW^*/d\vartheta} \cdot \left(\frac{G}{L_\alpha}\right) \cdot \left(1 + \frac{G \cdot t}{L_\alpha}\right)^2 \qquad (3.4.46)$$

or, after integration of Equation (3.4.46), to the equation of the cooling curve:

$$\frac{\vartheta_\alpha - \vartheta_t}{\vartheta_\alpha - \vartheta_\omega} = \frac{3}{M_{H_2O} \cdot \vartheta_\alpha} \cdot \frac{G \cdot t}{L_\alpha} \cdot \frac{M_\alpha}{dW^*/d\vartheta} \cdot \left[1 + \frac{G \cdot t}{L_\alpha} + \frac{1}{3} \cdot \left(\frac{G \cdot t}{L_\alpha}\right)^2\right] \qquad (3.4.47)$$

which can be approximated by the maximum term

$$\frac{\vartheta_\alpha - \vartheta_t}{\vartheta_\alpha - \vartheta_\omega} = \left(\frac{t}{t_c}\right)^3 \qquad (3.4.48)$$

where t_c is the batch or crystallization time.

According to Equations (3.4.47) and (3.4.48), cooling should initially be slow and only after the total crystal surface area has increased significantly should the cooling rate be increased. The linear crystal growth rate is then[74]:

$$G = 4^{1/3} \cdot \frac{(L_\omega - L_\alpha)}{t_c}$$

(3.4.49)

Essential to this technique is the value of supersaturation, ΔW, that must be measured and maintained in order to characterize the growth kinetics.

3.4.7 Simultaneous determination of G and $B_{o,eff}$ from batch experiments

In principle the rates of both crystal growth and nucleation can be determined from batch experiments. In a method described by Nývlt[74] the kinetics of crystal growth and nucleation depending on supersaturation were determined. During the operation of a seeded batch crystallization the super-saturation was monitored refractometrically and at the end of the experiment the crystal size distribution was measured. The crystal size distribution during the experiment and its dependence on time is then calculated iteratively. Taking the temperature dependence of crystal growth into account it is possible to determine growth and nucleation kinetics with the help of these results and the measured supersaturation. The experiments showed good results, for the determination of growth kinetics, but the results obtained from the calculation of nucleation rate varied widely from nucleation rates found in the literature.

Another method, proposed by Gutwald[76], recommends that constant super-saturation is maintained during the batch experiments in order to obtain kinetic data which are the same as would be determined from continuously operated crystallizers. The measurements of the concentration and the supersaturation have been achieved by a density meter fed by solution via a hydrocyclone which retained crystals larger than 10 μm. The change of the suspension density, m_T, with time can be drawn from a mass balance since the concentration as a function of time is known. It is also necessary to measure the crystal size distribution. It is known that the mean growth rate of crystals and especially of attrition fragments increases with increasing particle size. In order to overcome the problem of a suitable definition of a mean growth rate it is assumed that particles with the size $L < L_{eff}$ do not grow and particles with $L > L_{eff}$ grow with the constant mean growth rate G, which is defined in such a way that the material balance is fulfilled.

In Table 3.4.3 the expressions of the mean growth rate G and the effective rate of attrition controlled secondary nucleation $B_{0,eff}/\varphi$, based on the

Table 3.4.3 Rate expressions for continuously and batch operated crystallizers

Rate	Continuous	Batch
Growth: $\bar{G}_{k,r;>L_{eff}}$	$= \dfrac{1}{3\tau} \bar{L}_{k,r;L>L_{eff}}$	$= \dfrac{1}{3} \dfrac{d(\ln m_{T,>L_{eff}})}{dt} \bar{L}_{k,r;>L_{eff}}$
Sec. Nucleation $\dfrac{B_{0,eff}}{\varphi}$	$= \dfrac{1}{6\alpha G^3 \tau^4}$	$= \dfrac{9}{2\alpha} \dfrac{\dfrac{d(\ln m_{T,>L_{eff}})}{dt}}{\bar{L}_{1,2;>L_{eff}}^3}$

volumetric crystal hold-up φ, are presented. It is necessary to determine the suspension density m_T and the particle size distribution as a function of the cooling time, see Figures 3.4.11 and 3.4.12 (see page 104), which are valid for seeded potash alum solutions cooled down in a $7.2\,dm^3$ stirred vessel with a stirrer speed of $s = 370$ min^{-1} $= 6.17$ s^{-1}. In order to avoid primary nucleation the solution in the vessel was seeded with 1.5 g seed crystals of approximately 150 μm in size. In Figure 3.4.13 (see page 104) the stirred vessel and the measuring devices are depicted.

It could be shown that rates obtained from batch experiments agreed well with data determined in continuously operated MSMPR crystallizers.

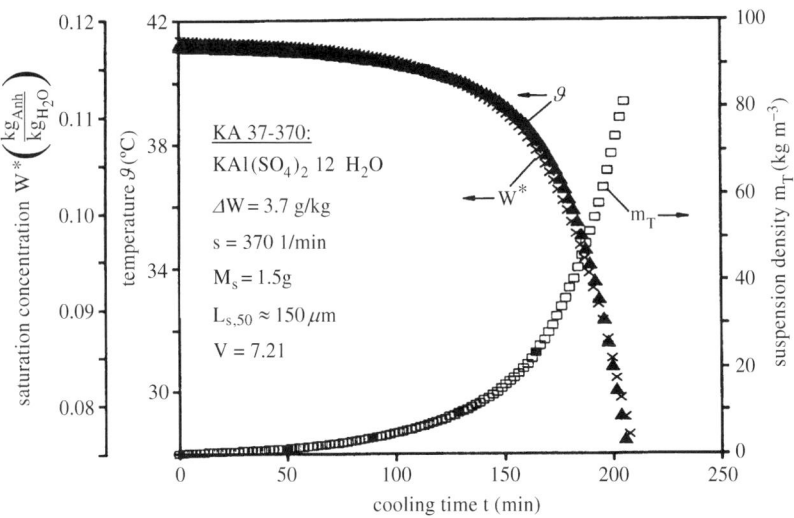

Figure 3.4.11 Evolution of conditions in batch cooling crystallization of potash alum[76]

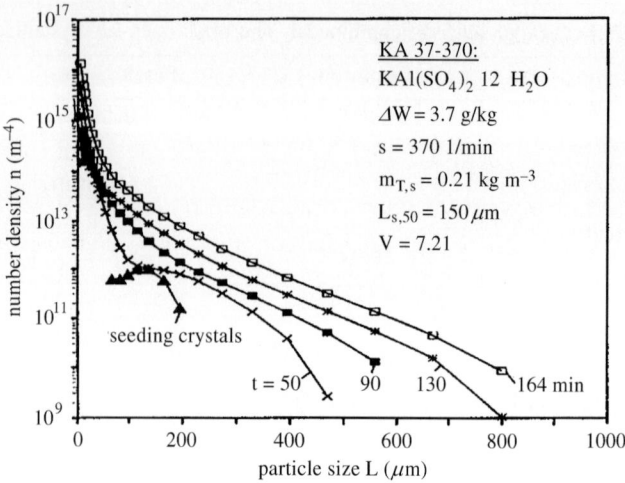

Figure 3.4.12 Development of size distribution during batch cooling crystallization of potash alum[76]

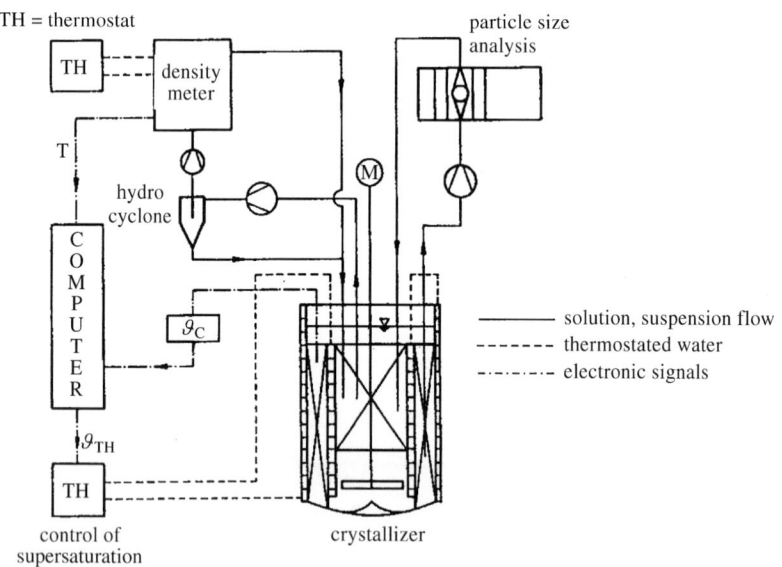

Figure 3.4.13 Equipment used for batch cooling experiments[76]

3.5 Continuous MSMPR crystallizer

A continuously operated Mixed Suspension Mixed Product Removal (MSMPR) crystallizer is an excellent device to determine the main kinetic parameters, crystal growth rate G and effective rate of nucleation B_0, which influence crystal size distribution.

In this section, information will be given on the measurement and analysis of crystal size distribution in order to obtain the crystal growth rate and effective nucleation rate.

3.5.1 MSMPR model

The MSMPR model will be briefly introduced and some limiting conditions will be pointed out in order to establish the basis for crystallization experiment analysis.

A comparison of the operating modes and the methods available to estimate kinetic parameters from measured crystal size distributions shows that the operation of mixed suspension mixed product removal crystallizers is most appropriate for this purpose. However, it is necessary to satisfy certain operating conditions, which impose limitations on the MSMPR method, and to apply recommended measuring and analysing procedures in order to get accurate results. An essential prerequisite is that the total crystallizer contents (solution and crystals) are perfectly mixed. Thus there must be no differences of concentration C or w, temperature ϑ, supersaturation ΔC or Δw, suspension density m_T and crystal size distribution within the suspension volume. Also, only solids with the same specification as that in the crystallizer must be removed. Further restrictions will be discussed later. As a result of these limitations, a small crystallizer $(3-20\,\mathrm{dm}^3)$ is recommended since it is easier to obtain good mixing in this case.

Figure 3.5.1 (see page 106) illustrates a continuously operated crystallizer. Supersaturation can be produced by cooling, evaporation or reaction. Steady state operation is assumed. Crystal-free feed solution is added and the slurry is removed isokinetically from the perfectly mixed suspension. In addition, processes such as attrition, breakage, agglomeration, dissolution and classification are assumed negligible and the crystallizer volume V is kept constant.

At steady state the population balance equation[77] becomes:

$$\frac{d[G(L) \cdot n(L)]}{dL} + \frac{n(L)}{\tau} = 0 \tag{3.5.1a}$$

105

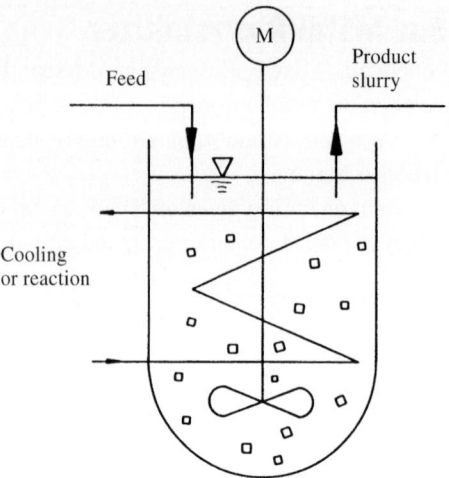

Figure 3.5.1 Continuously operated crystallizer

If crystal growth rate is size independent (i.e. the so-called ΔL law holds) this can be written:

$$G \cdot \frac{dn(L)}{dL} + \frac{n(L)}{\tau} = 0 \qquad (3.5.1b)$$

The assumptions noted above which lead to this simple equation will be termed 'idealized MSMPR conditions'. It is not always possible to accomplish these conditions; deviations from the idealized MSMPR model will be discussed later.

Integration of Equation (3.5.1b) gives the following relationship between the population density n, crystal size L, crystal growth rate G and residence time τ:

$$\frac{n(L)}{n_0} = \exp\left(-\frac{L}{G \cdot \tau}\right) \qquad (3.5.2)$$

or

$$\ln\left(\frac{n(L)}{n_0}\right) = -\frac{L}{G \cdot \tau}$$

Equation (3.5.2) gives a straight line when plotted on semi-logarithmic coordinates. An example of such a plot is given in Figure 3.5.2a (see page 107).

The crystal growth rate G can be deduced from the slope of the line which is equal to $-1/(G\tau)$, while the intersection of the straight line with the ordinate

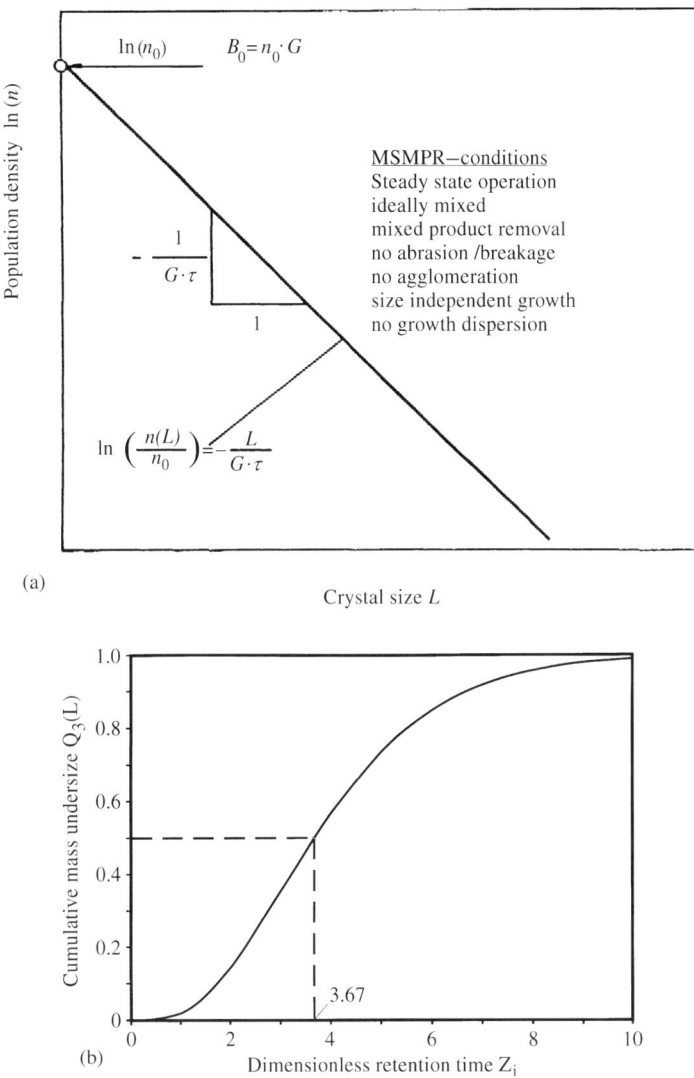

(a)

Crystal size L

(b)

Figure 3.5.2 (a) Population density plot and (b) cumulative mass undersize versus retention time

($L = 0$) gives the value $\ln(n_0)$. The product $n_0 \cdot G$ is the nucleation rate B_0, therefore:

$$B_0 = n_0 \cdot G \tag{3.5.3}$$

and hence the intercept: $n_0 = B_0/G$.

For a given residence time τ the crystal size distribution is thus completely described by the two kinetic parameters G and n_0 or G and B_0.

The moments of the size distribution can be calculated for the exponential number density distribution (Equation (3.5.2)) obtained with the idealized MSMPR model. They are shown in Table 3.5.1.

The nucleation rate B_0 based on the volumetric hold up φ is given by Equation (3.5.4) as a relationship between the crystal growth rate G and the residence time τ or the median diameter L_{50}:

$$\frac{B_0}{\varphi} = \frac{1}{6 \cdot \alpha \cdot G^3 \cdot \tau^4} = \frac{3.67^3}{6 \cdot \alpha \cdot L_{50}^3 \cdot \tau} = \frac{3.67^4 \cdot G}{6 \cdot \alpha \cdot L_{50}^4} \qquad (3.5.4)$$

or

$$L_{50} = 3.67 \cdot \sqrt[4]{\frac{G}{6 \cdot \alpha \cdot B_0/\varphi}}$$

This is only valid for the idealized MSMPR-model.

Crystal size distribution is carried out by sieve analysis as a rule. The cumulative mass undersize $Q_3(L)$ can be described by:

$$Q_3(L) = 1 - \frac{1 + Z_i + (Z_i^2/2) + (Z_i^3/6)}{\exp(Z_i)} \qquad (3.5.5)$$

Table 3.5.1 Moments of the population density distribution

$m_0 = \displaystyle\int_0^\infty n(L)dL$	Total number of crystals per unit suspension– volume $N_{tot} = m_0$	$N_{tot} = n_0 G\tau = B_0\tau$
$m_2 = \displaystyle\int_0^\infty L^2 n(L)dL$	Total surface area per unit suspension volume $a_T = \beta m_2$	$a_T = 2\beta n_0(G\tau)^3$
$m_3 = \displaystyle\int_0^\infty L^3 n(L)dL$	Crystal volume per unit suspension volume, $\varphi = \alpha m_3$, suspension density, $m_T = \alpha \rho_C m_3$	$\varphi = 6\alpha n_0(G\tau)^4$
		$m_T = 6\alpha \rho_C n_0(G\tau)^4$
		$m_T = \rho_C \varphi$
$\dfrac{m_3(0, L_{50})}{m_3} = 0.5$	Mass median particle size L_{50}	$L_{50} = 3.67 G\tau$
$0.5 = \dfrac{\displaystyle\int_0^{L_{50}} L^3 n(L)dL}{m_3}$		

In Figure 3.5.2b the expression $Q_3(L)$ is plotted against the dimensionless retention time $Z_i = L_i/(G\tau)$. For this distribution we obtain $Q_{3,15} = 0.556 \, Q_{3,50}$ and $Q_{3,85} = 1.638 \, Q_{3,50}$.

The median diameter L_{50} is defined in Table 3.5.1. This relationship is shown in Figure 3.5.3 for a volume shape factor $\alpha = \pi/6$. As will be shown later, it is very useful to plot experimental results in such a diagram.

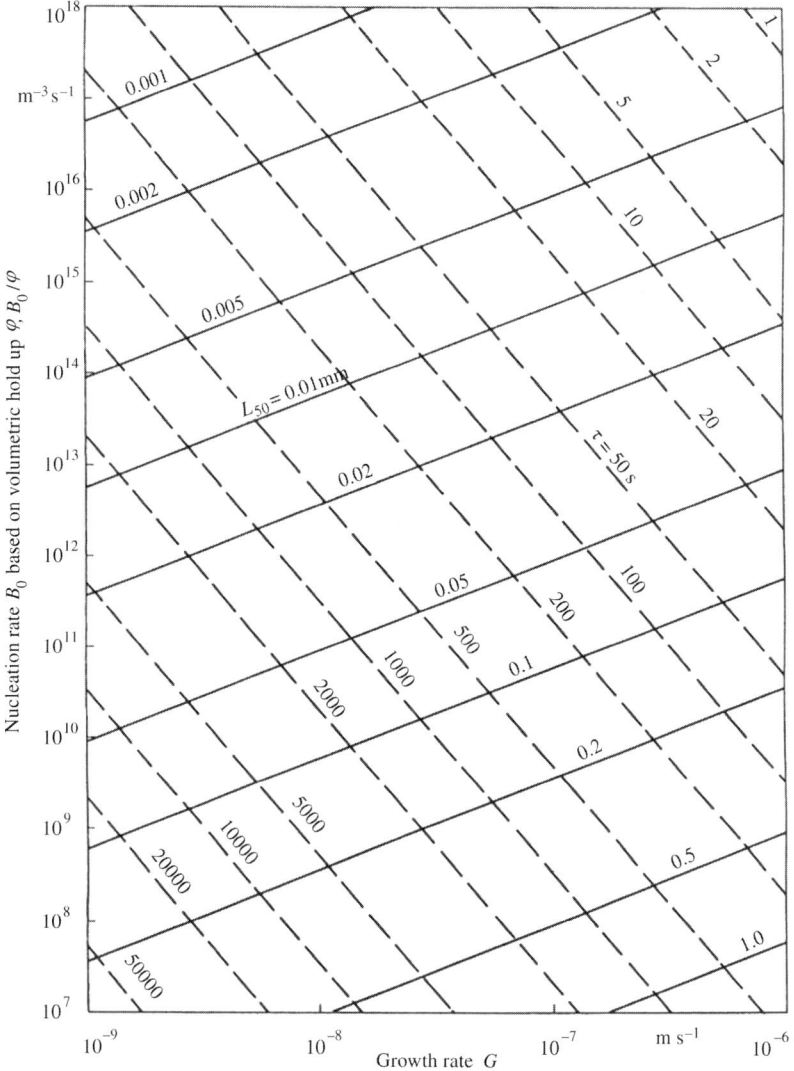

Figure 3.5.3 Idealized MSMPR model ($\alpha = \pi/6$ valid for spheres)

109

3.5.2 Measuring technique and analysis procedure

(a) Equipment for continuously operating MSMPR crystallizers
Recommendations for the flow sheet of the whole test plant are presented, followed by details of the MSMPR crystallizer.

Flow sheet
The flow sheet of a continuously operating MSMPR cooling crystallizer is shown in Figure 3.5.4. The process can be operated in a closed loop in order to obtain a relatively small total equipment volume. This, however, demands highly non-corrosive material for the crystallizer, tanks, tubes and pipes such as PVC, PMMA or stainless steel, otherwise the solution may become contaminated with corrosion products.

In order to maintain low consumption of crystalline material, the slurry withdrawal from the crystallizer is dissolved in a dissolution tank which is maintained at constant temperature by a thermostat TH. The crystal-free overflow of the dissolution tank goes to the storage tank and is then pumped through a filter to the head tank. The filter (opening $\approx 1\ \mu m$) holds back dust and any other solids which might influence the crystallization.

Most of the solution stream flows back to the dissolution and storage tanks in order to obtain good mixing between these tanks. A smaller part forms the

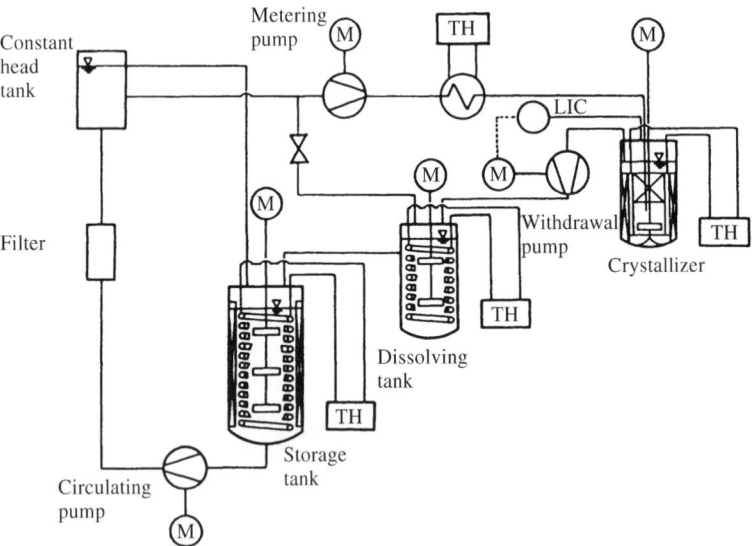

Figure 3.5.4 Flow sheet of a continuously operated crystallizer

110

feed to the crystallizer. The constant fluid level in the head tank provides a constant pressure in the suction line of the metering pump, which is necessary to obtain a constant flow rate to the crystallizer. A gravity feed and a flow meter to adjust the feed stream can also be used.

The feed can be cooled close to the saturation temperature in the heat exchanger before entering the crystallizer.

The crystallizer volume is a compromise between conflicting demands. A large volume is an advantage in obtaining representative results for the rate of secondary nucleation, particularly for the purpose of scale up. However it is expensive to build and operate a test plant with a large crystallizer volume. On the other hand a small crystallizer leads to low feed flow rates with the result that the pipe diameters are small. The possibility of plugging increases, if the pipe diameter is not small enough to ensure a sufficient velocity. Therefore a crystallizer volume of about 3 to $20\,dm^3$ is recommended for a good working test plant. For a cooling crystallizer a volume of about 3 to $10\,dm^3$ is preferable. For example a crystallizer volume $V = 6\,dm^3$, a flow rate $\dot{V}_f = 6\,dm^3\,h^{-1}$ and therefore a residence time $\tau = V/\dot{V}_f = 1\,h$ gives a velocity of $21\,mm\,s^{-1}$ in a pipe of $10\,mm$ diameter.

An evaporative crystallizer should have a volume of $20\,dm^3$ or more because in a smaller crystallizer the bubbles influence the fluid dynamics too much. It should be mentioned here that a coarser product can be obtained in larger crystallizers according to some scale up rules[12,78].

Steady state operation is usually obtained after operating for about 8 to 15 residence times. This demands a solution volume of up to $90\,dm^3$ for a crystallizer volume of $V = 6\,dm^3$. Since it is important to feed the crystallizer with exactly the same solution, free of crystals and having exactly the same temperature and concentration throughout the experiment, a solution volume of about $100\,dm^3$ would be needed for one experiment. Since this volume is rather large it is usually sufficient to recirculate the solution so long as it does not change composition during reheating, as illustrated in Figure 3.5.4 (see page 110). In the case of a $6\,dm^3$ crystallizer the total solution volume of the test plant is then approximately $60\,dm^3$. The volume of the dissolution tank should be about three times and the volume of the storage tank about five times the crystallizer volume.

It is recommended that the temperature in the dissolution tank and the storage tank are held at about $10\,K$ and $5\,K$ respectively above the solution saturation temperature. The solution flowing from the head tank to the crystallizer is cooled in the heat exchanger to a temperature approximately $0.5-2\,K$ above the saturation temperature. Thereby temperature profiles in the crystallizer can be minimized.

It is important to avoid classification of crystals during product removal. Therefore the slurry must be withdrawn isokinetically with the local bulk velocity of the suspension, a velocity that must be much higher than the settling velocity of the largest crystals. Product removal is best carried out intermittently using a peristaltic pump with an adjustable drive. In the case of an evaporative crystallizer, a piston pump may be preferable. Such withdrawal pumps operating intermittently are regulated by a level control loop.

The crystallizer
In order to obtain an ideally mixed crystallizer, a stirred vessel fitted with a draft tube and baffles (DTB) in the inner and outer area is most suitable. Such a crystallizer is shown in Figure 3.5.5. The draft tube itself is hollow and used as a heat exchanger to add or remove heat. The tube provides a stratified flow within the vessel with the result that the crystals in the slurry are distributed evenly in the whole crystallizer. The bottom of the crystallizer has a profiled shape in order to avoid settling of crystals and to ease crystal suspension. Experiments have shown that crystals can be suspended and distributed

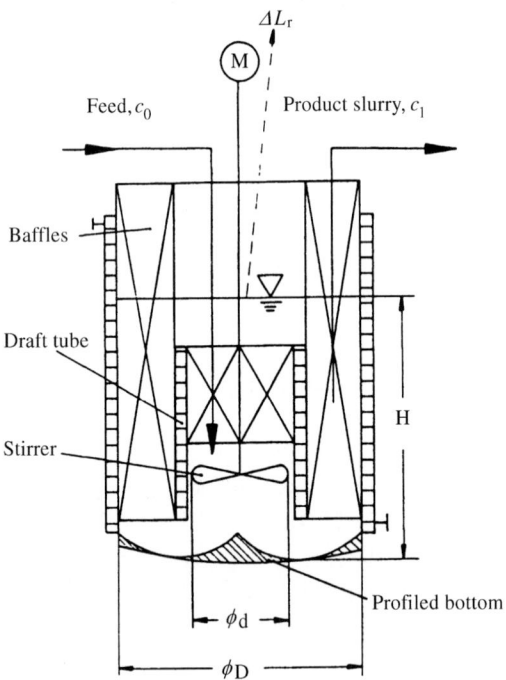

Figure 3.5.5 Draft tube baffled crystallizer (DTB)

uniformly at much lower stirrer speeds compared to a vessel without a draft tube and without a profiled bottom[79]. In addition, short circuiting is reduced with the result that attrition of crystals is reduced.

Another advantage of the stratified flow is that representative slurry removal is easier to obtain.

The crystallizer geometry is such that a height to diameter ratio of unity is often chosen for cooling and a ratio of about 2 for evaporation crystallizers. In order to avoid acceleration and deceleration of the slurry the cross-sectional areas of the inner tube, the outer annulus and between the draft tube and the profiled bottom should be the same. In this case the ratio of the stirrer to the vessel diameter d/D is between 0.6 and 0.67. The clearance between the stirrer and the inner wall of the draft tube should be at least 5 mm in order to avoid a milling effect of the stirrer on the crystals.

In order to obtain representative product removal the off-take tube should be at a height corresponding to $1/2$ to $2/3$ of the vessel diameter D shown in Figure 3.5.6. Above this position there might be an influence of the surface and below this region a certain degree of slurry segregation may take place. Such segregation arises from the inertial forces on the crystals as a consequence of the diversion of the flow within the loop so that solid material accumulates in

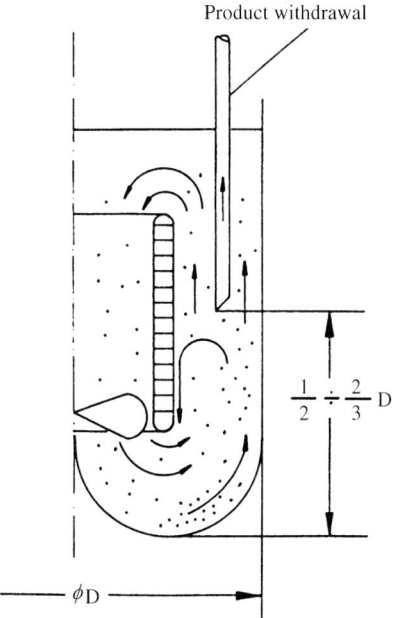

Figure 3.5.6 Position of the off-take tube

the vicinity of the vessel wall. Sometimes a back flow is also observed near the lower part of the draft tube as illustrated in Figure 3.5.6 on page 113.

During product removal the velocity within the withdrawal tube should be the same as the suspension velocity at that position, or about up to 20% higher than the velocity measured in solid free water. The flow rate in the vessel can be calculated by Equation (3.5.7).

The influence of classification and wash out experiments necessary to check the product removal will be discussed later.

If less than 10% of the whole vessel contents are removed, changes in the continuous steady state operating conditions of the crystallizer are relatively minor and can be ignored.

Since experimental results obtained in MSMPR crystallizers are used to establish nucleation models for the design and scale-up of industrial crystallizers, it is very important to give sufficient information on the crystallizer used. A minimum of the following data are needed:

crystallizer volume	V
crystallizer diameter	D
filling height	H
inner and outer draft tube diameters	$D_{D,i}$ and $D_{D,a}$
stirrer diameter	d
stirrer speed	s
number of blades	a
blade thickness	t
blade width	b
blade angle	β
or	
stirrer slope	
material of construction of stirrer	
type of stirrer	

Even if the further use of some of these data is not explained here explicitly, they can be important for evaluating the kinetic parameters like the rate of secondary nucleation.

It is recommended that the Power number P_o and the flow number N_V are determined experimentally. These numbers are defined by the expressions:

Stirrer power $\quad P = P_o \cdot \rho_{sus} \cdot s^3 \cdot d^5$ (3.5.6)

Mean specific power input $\quad \bar{\varepsilon} = \dfrac{P_o \cdot s^3 \cdot d^5}{V}$ (3.5.6a)

Pumping rate $\quad \dot{V} = N_V \cdot s \cdot d^3$ $\hspace{6cm}$ (3.5.7)

At high stirrer Reynolds numbers the value of the Power number P_o for a marine type propeller in a DTB crystallizer is between about 0.3 and 0.55 and the flow number N_V about 0.3 to 0.45.

In principle a fluidized bed without impeller and with only crystal–crystal collisions can also serve as a crystallizer for MSMPR-experiments[52], especially if the mixedness of the crystallizer's contents does not influence the size distribution very much. The difficulty of removing and sampling the crystals *representatively* can be overcome by using an off-take tube driven through the crystallizer according to the local cross-sectional area, velocity distribution and solid concentration in the fluidized bed. The sampling should be checked experimentally by wash-out experiments or compared with the crystallizer contents.

(b) Experimental method

After reaching steady state, which is usually obtained after a period of 8 to 15 residence times, product can be used to determine the crystal size distribution and the kinetic parameters, crystal growth rate and nucleation rate.

Depending on the median crystal size and the operating conditions of the crystallizer, various methods for the measurement of crystal size distribution are available (compare Section 2.7). For the size range above $100\,\mu$m wet and/or dry sieving is usually applied. In the case of a low viscosity solution, crystallized at room temperature, the suspension can be sieved directly using a wet sieve shaker with a filter following the finest sieve. If the temperature of the slurry differs markedly from room temperature, the resulting change in temperature would lead to further crystallization or dissolution with the result that the original distribution is not determined.

Sometimes slurry sieving of viscous suspensions may be troublesome. In these cases the crystals can be separated from the suspension by means of a filter and then washed with an inert liquid which does not dissolve the crystalline material. After filtration and drying, the crystal size distribution can be determined down to $100\,\mu$m by dry sieving.

The shape factors of the crystals have to be taken into account. Definition and measurement techniques of shape factors are described in Section 2.6.

It should be emphasized that any change of the crystal size distribution during the various procedures of filtering, washing or sieving must be avoided or reduced to a minimum, otherwise errors can occur by further crystallization or dissolution of material as well as by agglomeration or breakage.

(c) Determination of crystal size distribution, growth rate and nucleation rate

The following example shows how the crystal size distribution can be determined by slurry sieving. Comments on the calculation of crystal growth rate and rate of nucleation are given.

Geometric and operating data of the crystallizer:

Crystallizer:	volume	$V = 5.7\,\text{dm}^3$
	diameter	$D = 0.2\,\text{m}$
	filling height	$H = 0.2\,\text{m}$
Draft tube:	inner diameter	$D_{D,i} = 0.11\,\text{m}$
	outer diameter	$D_{D,a} = 0.13\,\text{m}$
Stirrer:	marine type propeller	
	diameter	$d = 0.1\,\text{m}$
	number of blades	$a = 3$
	blade thickness	$t = 0.0015\,\text{m}$
	blade width	$b = 0.033\,\text{m}$
	blade angle	$\beta = 25°$
	Power number	$P_o = 0.36$
	Flow number	$N_V = 0.30$

Operating parameters: (obtained for the $KNO_3 - H_2O$ system)

feed flow rate	$\dot{V}_f = 11.9\,\text{dm}^3\,\text{h}^{-1}$
residence time	$\tau = 1730\,\text{s}$
feed temperature	$\vartheta_f = 35.6°C$
feed equilibrium temperature	$\vartheta_f^* = 34.1°C$
crystallizer temperature	$\vartheta_C = 22.3°C$
room temperature	$\vartheta_R = 22.3°C$
stirrer speed	$s = 7.5\,\text{s}^{-1}$
mean specific power input	$\bar{\varepsilon} = 0.27\,\text{W}\,\text{kg}^{-1}$

A slurry sieve shaker was used. The sieve meshes increased by the factor $(2)^{1/4}$ in the size range $150\,\mu\text{m} < L < 1000\,\mu\text{m}$ and by the factor $(2)^{1/2}$ above $1000\,\mu\text{m}$. The sieve with the smallest opening was provided with a filter paper which retained nearly all the fine product.

In order to improve the accuracy of the CSD determination, a number of crystals taken from different sieves with the width L_s were counted and weighed to calculate the mass median crystal size L_{50} on every sieve. It was assumed that the shape factor $\alpha = \pi/6$ is not a function of the crystal size (see Table 3.5.2 on page 118). The size of the sieve opening L_s was corrected using the actual sizes of particles on each sieve.

116

The crystalline mass on a single sieve, M_C, was calculated from

$$M_C = M_{dry} - M_{sieve} - (M_{wet} - M_{dry}) \cdot W^*(\vartheta_C) \qquad (3.5.8)$$

with:

$M_{sieve} =$ mass of the clean, dry sieve

$M_{wet} =$ mass of wet crystals and sieve

$M_{dry} =$ mass of dry crystals and sieve

$W^*(\vartheta_C) =$ equilibrium concentration of the solution

(kg/kg H_2O at crystallizer temperature ϑ_C).

The expression $(M_{wet} - M_{dry}) \cdot W^*(\vartheta_C)$ takes into account the crystalline mass which crystallizes while drying the wet product. This term may be as large as 30% of M_C and more for highly soluble systems such as KNO_3 in water with a solubility of about 0.35 kg/kg H_2O at room temperature.

For a given size interval ΔL the mass fraction per unit interval is given by

$$q_3(\bar{L}_i) = \frac{M_i}{\sum_{j=1}^{T} M_j} \cdot \frac{1}{\Delta L_i} \qquad (3.5.9)$$

where T is the total number of the sieves in the shaker.

The cumulative mass undersize can be calculated from:

$$Q_3(L_i) = \frac{\sum_{j=1}^{i} M_j}{\sum_{j=1}^{T} M_j} \qquad (3.5.10)$$

The population density n, i.e. the number of crystals in an interval ΔL per unit suspension volume, can be expressed by:

$$n(\bar{L}_i) = \frac{M_i}{V_{sample} \cdot \alpha \cdot L_i^3 \cdot \rho_C \cdot \Delta L_i} \qquad (3.5.11)$$

Values for $q_3(L_i)$, $Q_3(L_i)$ and $n(\bar{L}_i)$ are given in Table 3.5.2 (see page 118) and Figures 3.5.7 and 3.5.8 (see pages 120 and 121).

The crystallizer suspension density m_T can be calculated as the sum of all masses M_i on the sieves referred to the sample volume V_{sample}.

$$m_T = \frac{\sum_{j=1}^{T} M_j}{V_{sample}} \qquad (3.5.12)$$

Table 3.5.2 Sieve analysis data

L_s[µm]	$L_{s,corr}$[µm]	L_{50}[µm]	ΔL[µm]	M_C[g]	q_3[m^{-1}]	Q_3[-]	$n(L)$[m^{-4}]	N/V[m^{-3}]	$A/V\beta$[m^2/m^{-3}]
0	0					0.000			
		85	170	1.90	110.6		2.1E13	3.6E9	25.73
150	170					0.019			
		185	30	1.36	448.5		8.2E12	2.5E8	8.46
180	200					0.032			
		215	30	1.66	547.4		6.4E12	1.9E8	8.89
212	230					0.049			
		250	40	3.29	813.7		6.1E12	2.4E8	15.15
250	270					0.081			
		295	50	5.60	1108.0		5.0E12	2.5E8	21.85
300	320					0.137			
		350	60	8.58	1414.7		3.8E12	2.3E8	28.22
355	380					0.222			
		420	80	12.49	1544.6		2.4E12	1.9E8	34.23
425	460					0.345			
		505	90	14.86	1633.5		1.5E12	1.3E8	33.88

500	550	600	100	21.81	2157.7	0.492	1.2E12	1.2E8	41.85
650	650	705	110	8.68	780.7	0.708	2.6E11	2.9E7	14.17
760	760	820	120	10.29	848.3	0.794	1.8E11	2.1E7	14.45
880	880	910	120	7.62	628.2	0.896	9.7E10	1.2E7	9.64
1000	1000	1150	300	2.52	83.1	0.971	6.4E09	1.9E6	2.52
1300	1300	1600	600	0.42	6.9	0.996	2.0E08	1.2E5	0.30
1900	1900					1.000			
Σ				101.08				5.2E9	259.35

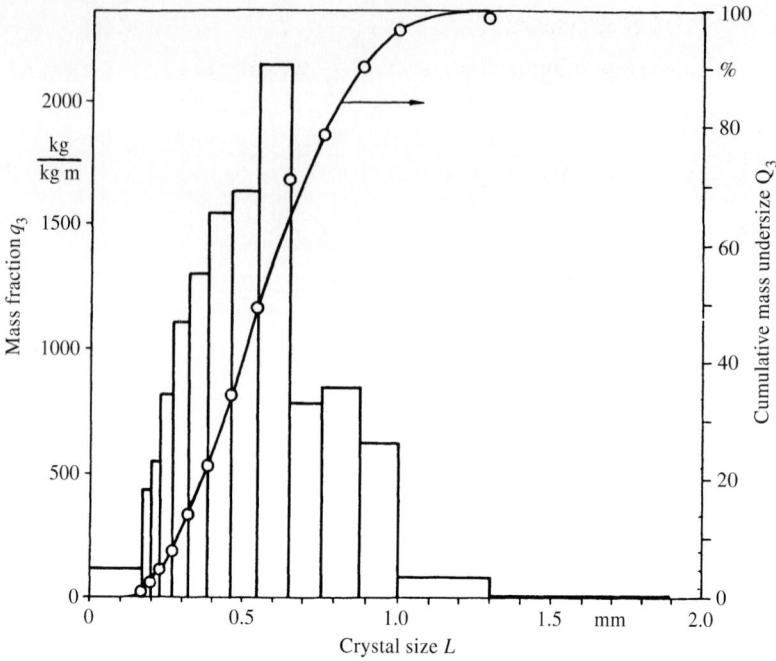

Figure 3.5.7   Mass fraction per unit interval versus crystal size and cumulative mass undersize according to Table 3.5.2

This suspension density should be compared with that calculated from the mass balance on the whole crystallizer (cooling or evaporating):

$$m_T = \frac{M_f \cdot \rho_f}{\dot{M}_1} - \rho_1 \cdot \left(1 - \frac{m_T}{\rho_C}\right) \tag{3.5.13}$$

or

$$m_T = \frac{\left[\frac{\dot{M}_f \cdot \rho_f}{\dot{M}_1} - \rho_1\right]}{\left(1 - \frac{\rho_1}{\rho_C}\right)}$$

with

$\dot{M}_f$ = feed mass flow rate [kg solution s^{-1}]

$\dot{M}_1$ = discharge mass flow rate [kg suspension s^{-1}]

ρ_C = mass density of crystals [kg m^{-3}]

120

C_f = feed concentration [$kg\,m^{-3}$ suspension]

C_1 = discharge solution concentration [$kg\,m^{-3}$ solution]

and

$$C_1 = C * (\vartheta_C) + \Delta C \qquad (3.5.14)$$

with

$C * (\vartheta_C)$ = saturation concentration of the crystallizer temperature

$\times$ [$kg\,m^{-3}$ solution]

ΔC = supersaturation [$kg\,m^{-3}$ solution]

Sometimes the volumetric fraction $\varphi = m_T/\rho_C$ is much smaller than unity. In this case ($\varphi \ll 1$) we obtain.

$$m_T = \frac{\dot{M}_f \cdot C_f}{\dot{M}_1} - C_1 \qquad (3.5.15)$$

A material balance over the whole crystallizer gives:

$$\dot{M}_f = \dot{M}_1 + \Delta\dot{M} \qquad (3.5.16)$$

where $\Delta\dot{M}$ is the evaporation mass rate. For a cooling crystallizer, evaporation does not take place and $\Delta\dot{M} = 0$.

Combining Equation (3.5.15) with (3.5.14) and (3.5.16) we obtain:

$$m_T = \frac{C_f}{1 - \Delta\dot{M}/\dot{M}_f} - C * (\vartheta_C) - \Delta C \qquad (3.5.17)$$

For a cooling crystallizer ($\Delta\dot{M} = 0$) when $\varphi \ll 1$ the last equation simplifies to

$$m_T = C_f - C_1 \qquad (3.5.18)$$

or with the additional condition $\Delta C \ll C_1$ to

$$m_T = C_f - C * (\vartheta_C) \qquad (3.5.19)$$

A comparison of the slurry densities obtained by the summation of the masses on the sieves [Equation (3.5.12)] and by calculation with the concentrations C_f and C_1 [Equation (3.5.13) and (3.5.18)] may indicate that a mistake has occurred.

The mean crystal growth rate G and the value of n_0 can be obtained from a plot ln n against L by applying a regression analysis. Figure 3.5.8 (see page 122) shows the straight line determined in this way. The rate of nucleation B_0 can be calculated with Equations (3.5.3) or (3.5.4).

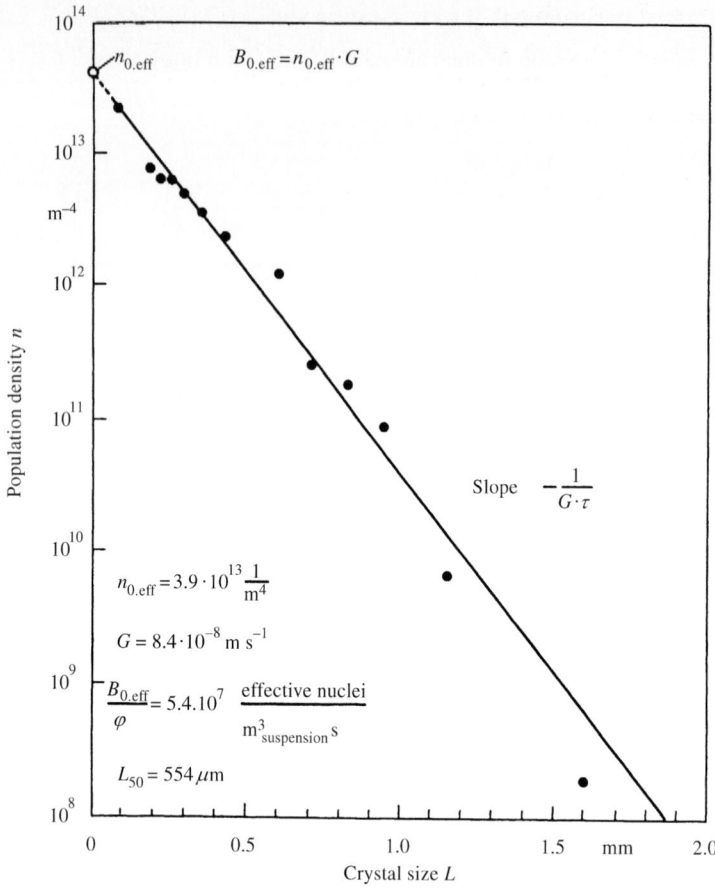

Figure 3.5.8 Population density n versus crystal size L according to Table 3.5.2

Often the data below $L \approx 100\,\mu m$ deviate from a straight line. This may be caused by size dependent growth and/or growth rate dispersion. Sometimes data above $L \approx 100\,\mu m$ also show deviations from the straight line due to severe attrition. In these cases it is recommended that the kinetic data for G and B_o is calculated from a straight line in the approximate range $100\,\mu m$ to $1000\,\mu m$.

Good straight lines on a $\ln n$ against L plot are seldom obtained when experiments are carried out for high suspension densities ($m_T >\approx 100\,kg\,m^{-3}$), high specific power inputs ($\bar{\varepsilon} >\approx 0.5\,W\,kg^{-1}$) and long residence times ($\tau >\approx 5000\,s$).

Such non-linearity is frequently caused by additional effects which may result mainly from the crystallizer (such as attrition, breakage, agglomeration and non-ideal mixing) and effects that are characteristic of the crystallized system (such as size dependent growth and growth rate dispersion). Describing such processes is difficult due to the lack of reliable models for these effects, but the median effective values for the growth rate $\bar{G}_{eff}$ and the rate of nucleation $B_{0,eff}$ may be calculated by applying the equations[78,149].

$$\bar{G}_{eff} = \frac{m_T}{3 \cdot \alpha \cdot \rho_C \cdot \tau \cdot (A/V)} = \frac{m_T}{3 \cdot \alpha \cdot \rho_C \cdot \tau \cdot \sum_{i=1}^{T} n_i(\bar{L}_i) \cdot \bar{L}_i^2 \cdot \Delta L_i} \quad (3.5.20)$$

$$B_{0,eff} = \frac{1}{\tau} \cdot \sum_{i=1}^{T} \left[\frac{N}{V}\right]_i = \frac{1}{\tau} \cdot \sum_{i=1}^{T} n_i(\bar{L}_i) \cdot \Delta L_i. \quad (3.5.21)$$

Values for the expression (A/V) and $(N/V)_i$ are included in Table 3.5.2 (see page 118).

The equation for the mean effective growth rate $\bar{G}_{eff}$ is based on a mass balance assuming that all crystals are growing with the same rate independent of their size. The value of $B_{0,eff}$ represents all particles that originate from effects other than crystal growth and are in the measured size range. At this point it should be pointed out that experimental values of the rate of nucleation mostly depend on the size of the smallest detectable particle. The extrapolation to zero-size particles in the semi-logarithmic population density plot may produce significantly different results if sieve analysis is compared with, for example, Coulter Counter[80] or Fraunhofer diffraction measurements. The intercept value of $n(L)$ for $L \to 0$ is termed $n_{0,eff}$, see Figure 3.5.8.

The rate of nucleation $B_{0,eff}$ is often divided by the volume fraction φ. Thus Equation (3.5.21) can be transformed to:

$$\frac{B_{0,eff}}{\varphi} = \frac{1}{\alpha \cdot \tau} \cdot \frac{\int_0^{L_{max}} n(L)\,dL}{\int_0^{L_{max}} L^3 \cdot n(L)\,dL} \quad (3.5.22)$$

with:

$$\varphi = \alpha \cdot \int_0^{L_{max}} L^3 \cdot n(L)\,dL \quad (3.5.23)$$

Table 3.5.3 (see page 124) shows a comparison of results obtained when the various equations and calculation methods presented earlier are applied to the above set of experimental data.

Within a range of accuracy sufficient for industrial purposes the kinetic data $\bar{G}_{eff}$ and $B_{0,eff}/\varphi$ calculated by the regression analysis are approximately the same as data according to a summation of the number or area per unit volume [Equations (3.5.20) and (3.5.21)]. The differences in the growth rate are about 11%, while in the nucleation the range is approximately 20%.

CRYSTAL GROWTH AND NUCLEATION RATES

Table 3.5.3 Results of sieve analysis for the example: (KNO$_3$: V$=5.7\,$dm^3; $\tau=1730$ s; s$=7.5$ s^{-1})

	From sieve analysis as in Equation (3.5.12) with $\varphi=m_T/\rho_C$	From mass balance as in Equation (3.5.13)	
φ [–]	0.061	0.056	
	From the slope of regression line	From mass balance as in Equation (3.5.20)	
$\bar{G}_{eff}$[m s^{-1}]	$7.8\cdot10^{-8}$	$8.7\cdot10^{-8}$	
$n_o=$[m^{-4}]	$4.8\cdot10^{13}$	/	
	From regression line as in Equation (3.5.4)	From the sum of particle numbers in Equation (3.5.21)	With growth rate G in Equation (3.5.20) and Equation (3.5.5)
$\alpha=\pi/6$ $B_{o,eff}/\varphi$[m^{-3}s^{-1}]	$6.2\cdot10^{-7}$	$5.0\cdot10^{7}$	$5.5\cdot10^{7}$
	From cumulative mass	With growth rate G from slope of regression line and $L_{50}=3.67\,G\tau$	With growth rate G as in Equation (3.5.20) and $L_{50}=3.67\,G\tau$
$\alpha=\pi/6$ L_{50} [μm]	554	494	563

124

It is recommended that sufficient information on measuring techniques and calculation methods for crystal size distribution is always given to enable researchers to compare their own results with those of other authors.

3.5.3 Estimation of the effective rate of attrition controlled secondary nucleation, $B_{0,eff}$

It is known that attrition fragments are the main source of nuclei when brittle crystals with a high solubility are produced at a low supersaturation. In this case activated nucleation is usually negligible in industrial crystallizers. Potassium nitrate is a brittle material and belongs to the systems with a high solubility $(c^* > 0.1\,\text{kmol m}^{-3})$.

Crystallizers for such systems are always operated in the range $0.001 < \sigma < 0.1$ or $1.001 < S < 1.1$ where rates of activated nucleation are extremely low. Therefore, nucleation is controlled by attrition fragments. Many experiments have shown that a huge number $N_{a,tot}$ of attrition fragments is produced in industrial crystallizers but only a small fraction of them has the ability to grow at these low supersaturations. Many authors have measured the size distribution of attrition fragments resulting from parent crystals suspended in a stirred vessel[81-85]. In Figure 3.5.9 the number density and the cummulative number distribution of KNO_3 attrition fragments are plotted against the attrition fragment size L_a according to Gahn[82].

It is rather surprising that only attrition fragments in the size range between $2\,\mu m$ and $200\,\mu m$ have been measured for stirrer tip speeds below $5\,\text{m s}^{-1}$. According to Gahn[82] the minimum size $L_{a,min}$ is given by:

$$L_{a,min} = 9.4\,\frac{\mu}{H_V^2}\left(\frac{\Gamma}{K_r}\right) \tag{3.5.24}$$

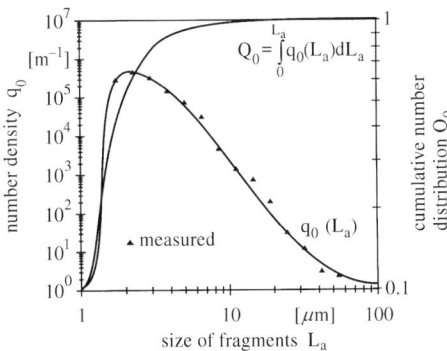

Figure 3.5.9 Number density and cummulative number distribution of KNO_3 attrition fragments[82]

125

and the maximum size $L_{a,max}$ can be calculated from:

$$L_{a,max} = 0.5 \frac{H_v^{2/9}}{\mu^{1/3}} \left(\frac{K_r}{\Gamma}\right)^{1/3} W_{col}^{4/9} \quad (3.5.25)$$

In these equations H_v is the Vickers Hardness, μ is the shear modulus, Γ is the fracture resistance, K_r is an efficiency factor and W_{col} is the collision energy which increases with the square of the collision velocity w_{col} according to:

$$W_{col} = \alpha L_{par}^3 \rho_C \frac{w_{col}^2}{2} \quad (3.5.26)$$

Here L_{par} is the size of the parent crystal and ρ_C is its density.

Theoretical considerations as well as many experimental results lead to the following equation for the total number $N_{a,tot}$ of attrition fragments resulting from one parent crystal which absorbs the collision energy W_{col}:

$$N_{a,tot} = 7 \cdot 10^{-4} \frac{H_v^5}{\alpha \mu^3} \left(\frac{K_r}{\Gamma}\right)^3 W_{col} = 7 \cdot 10^{-4} \frac{H_v^5}{\mu^3} \left(\frac{K_r}{\Gamma}\right)^3 L_{par}^3 \frac{\rho_C W_{col}^2}{2}$$

$$(3.5.27)$$

It is important to note that nearly all small attrition fragments possess an increased chemical potential (elevated solubility) due to deformation and do not grow at a low supersaturation[86,87]. Therefore, the number $N_{a,eff}$ of effective attrition fragments is decisive for the rate of secondary nucleation. The effectiveness of KNO_3 attrition fragments has been determined by Wang[85], see Figure 3.5.10 on page 127. Here the growth rate of 356 fragments has been measured in growth cells.

As can be seen only fragments larger than about $L_{a,eff}=25\,\mu m$ contribute significantly to growth $(G(L_a) > 0.1\bar{G}_{max})$. In the Figures 3.5.9 and 3.5.11 on page 127 the ratio $N_{a,eff}/N_{a,tot}$ according to:

$$\frac{N_{a,eff}}{N_{a,tot}} = \int_{25\mu m}^{\infty} q_0(L_a)\,dL_a = 1 - \int_{0}^{25\mu m} q_0(L_a)\,dL_a \quad (3.5.28)$$

is $N_{a,eff}/N_{a,tot} \approx 0.017$.

It is important to mention that this ratio is only valid for the supersaturation $\Delta c=0.02\,mol\,l^{-1}$ and rises with increasing supersaturation.

In the following a very simplified model will be presented in order to show how it is possible to give an estimation of the effective rate of attrition controlled secondary nucleation, $B_{0,eff}$. The purpose of this crude estimation is to make clear the importance of the material properties H_v, μ and (Γ/K_r) which are described in more detail in Section 3.6.

126

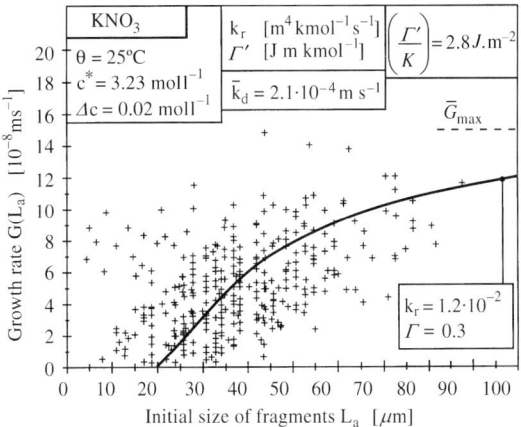

Figure 3.5.10 Growth rate of 356 KNO_3 attrition fragments[85]

The mean specific power input, $\bar{\varepsilon}$, can be expressed by the power number P_o, the stirrer diameter d and the stirrer tip speed $u_{tip} = d \pi s$:

$$\bar{\varepsilon} = \frac{P_o}{\pi^3} \cdot \frac{u_{tip}^3 d^2}{V_{sus}} \qquad (3.5.29)$$

Here, V_{sus} is the volume of the suspension. The collision velocity, w_{col}, is smaller than the tip speed:

$$w_{col} = \eta_w u_{tip} \quad \text{with } \eta_w < 1 \qquad (3.5.30)$$

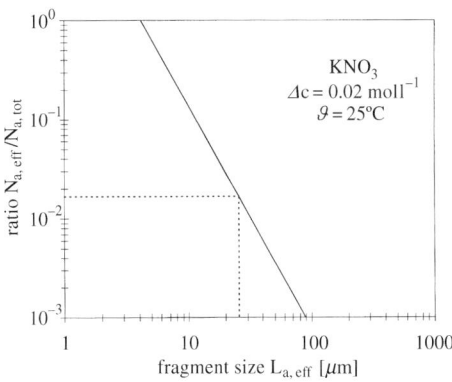

Figure 3.5.11 Ratio $N_{a,eff}/N_{a,tot}$ as a function of the size $L_{a,eff}$ of effective attrition fragments according to Equation (3.5.28)

127

and the frequency of contacts, f_c, of a parent crystal with the rotor is smaller than the speed s:

$$f_c \approx \eta_g \eta_w \frac{\dot{V}}{V_{sus}} = \eta_g \eta_w \frac{N_V s d^3}{V_{sus}} \tag{3.5.31}$$

with the pumping capacity N_V and η_g (for marine type propeller) ≈ 0.03. A combination of the Equations (3.5.27), (3.5.29), (3.5.30) and (3.5.31) leads to

$$\frac{f_c N_{a,tot}}{V_{sus}} = 7 \cdot 10^{-4} \frac{H_V^5}{\mu^3} \left(\frac{K_r}{\Gamma}\right)^3 \frac{\pi^2 \rho_C \bar{\varepsilon} N_V}{2 P_o} \frac{L_{par}^3}{V_{sus}} \eta_w^3 \eta_g \tag{3.5.32}$$

The right hand side of this equation has to be multiplied by the factor $N_{a,eff}/N_{a,tot}$ and by the total number of parent crystals which amounts to $V_{sus}\varphi/(\alpha L_{par})^3$ in the case of monodispersed parent crystals with the shape factor α. This leads to the effective rate of secondary nucleation based on the volumetric crystal hold-up:

$$\frac{B_{0,eff}}{\varphi} = 7 \cdot 10^{-4} \frac{H_V^5}{\mu^3} \left(\frac{K_r}{\Gamma}\right)^3 \frac{\pi^2 \rho_C \bar{\varepsilon} N_V}{2 \alpha^3 P_o} \frac{N_{a,eff}}{N_{a,tot}} \eta_w^3 \eta_g \tag{3.5.33}$$

With
$$H_V = 2.68 \cdot 10^8 \, N\,m^{-2}$$
$$\mu = 7.17 \cdot 10^9 \, N\,m^{-2}$$
$$\left(\frac{\Gamma}{K_r}\right) = 2.8 \, J\,m^{-2}$$
$$\rho_C = 2109 \, kg\,m^{-3}$$
$$\bar{\varepsilon} = 0.27 \, W\,kg^{-1}$$
$$P_o = 0.36$$
$$\eta_w = 0.8^{82,84}$$
$$\eta_g = 0.03$$
$$N_V = 0.3$$
$$\alpha = 1$$
$$\frac{N_{a,eff}}{N_{a,tot}} = 0.017$$

the result is $B_{0,eff}/\varphi \approx 7.4 \cdot 10^7$ nuclei/m^3 s. In Section 3.6 it will be shown how the material properties hardness H_V, shear modulus μ and fracture resistance Γ/K_r can be determined experimentally by simple indentation tests.

However, the accuracy of such data obtained experimentally and/or theoretically is not very high since the solid properties can differ from crystal to crystal. The main drawback of the model presented above is that the ratio

$N_{a,eff}/N_{a,tot}$ is unknown and can only be obtained from the growth behaviour of a large number of attrition fragments.

3.5.4 Comparison of crystal growth rates obtained for single crystals, for crystals in fluidized beds and from MSMPR crystallizers

Investment and operation costs of continuously operated MSMPR crystallizers are high. Experiments carried out on small fluidized beds, and especially in a single crystal cell, in order to determine crystal growth rates are much less expensive. The question therefore arises whether growth rates obtained in these apparatus are reliable for the design of industrial crystallizers. It is difficult to give a general answer, but the limitations of experiments in a single crystal cell are mainly as follows:

- Only the growth of a single face is usually observed and this is unlikely to be representative of the mean growth rates of all faces. In fluidized beds and especially in stirred vessels the effect of attrition often results in the larger crystals being rounded without showing distinct faces.
- Small amounts of impurities can reduce the crystal growth rates of a small number of crystals to a high degree since their surface area is so small. In fluidized beds and stirred vessels this may be of minor importance due to the large crystal surface area. This effect was observed for KCl in the presence of $PbCl_4$[51,52,147] and for NaCl with small concentrations of ferrocyanate present[89].
- Information on the influence of mechanical treatment on crystal size distribution and the rate of secondary nucleation can be obtained for fluidized beds and stirred vessels. However, the effects of agglomeration and destruction of agglomerates are different in single crystal cells, fluidized beds and stirred vessels due to the different fluid dynamic properties (ratios $\varepsilon/\bar{\varepsilon}$ of local to mean specific power input as high as 15 and as low as 0.2 occur according to Geisler and Mersmann[88]).
- If crystal growth is influenced by bulk diffusion, growth rate experiments should be carried out in such a way that either the mass transfer coefficients are known or can be determined with sufficient accuracy. When mass transfer effects can be calculated, fairly good agreement of growth rates determined in single crystal cells, fluidized beds and stirred vessels can be observed. This is illustrated in Figure 3.5.12 (see page 130) for the systems KNO_3, KCl, $(NH_4)_2SO_4/H_2O$ at high supersaturations $\Delta C/\Delta C_{met} \approx 0.5$ where ΔC_{met} is the metastable supersaturation.

In Figure 3.5.12 the dimensionless growth rates $G/(2k_d)$ are plotted against the dimensionless supersaturation $\Delta C/\rho_c$ for data measured in a single crystal cell,

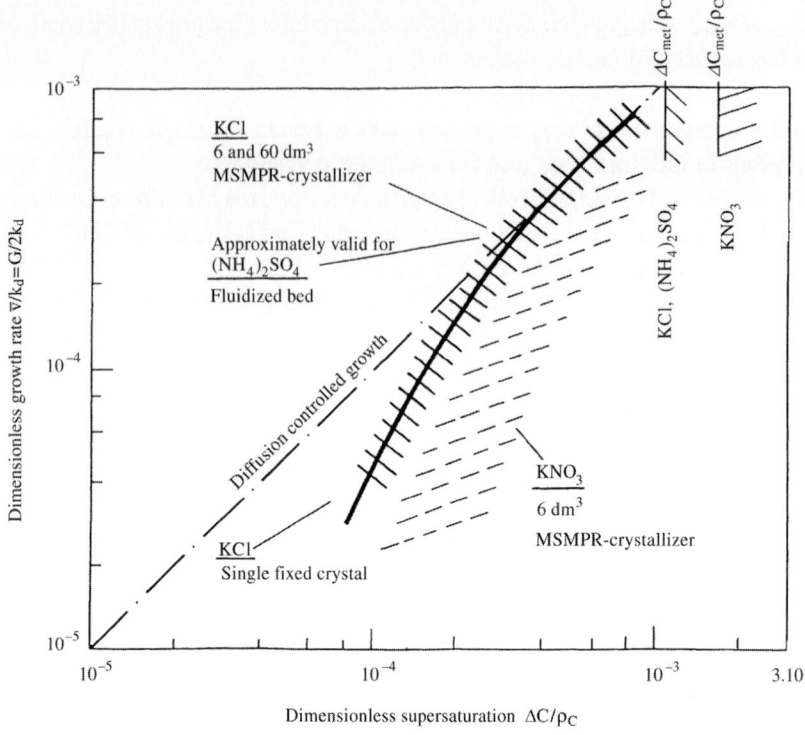

Figure 3.5.12 Comparison of growth rates obtained from experiments with single crystals, in fluidized beds and in MSMPR crystallizers

in a fluidized bed with about 15 suspended crystals and in a $5.7\,\text{dm}^3$ MSMPR crystallizer[13,90]. The mass transfer coefficients were determined by the relationships[91]:

Single sphere:

$$\text{Sh}_{\text{sp}} = \frac{k_d \cdot L}{D_{AB}} = 2 + \sqrt{\text{Sh}_{\text{lam}}^2 + \text{Sh}_{\text{turb}}^2} \qquad (3.5.34)$$

with:

$$\text{Sh}_{\text{lam}} = 0.664 \cdot \sqrt{\text{Re}_\psi} \cdot (\text{Sc})^{1/3}$$

$$\text{Sh}_{\text{turb}} = \frac{0.037 \cdot \text{Re}_\psi^{0.8} \cdot \text{Sc}}{1 + 2.44 \cdot \text{Re}_\psi^{-0.1} \cdot (\text{Sc}^{2/3} - 1)}$$

and the bed Reynolds number

$$Re_\psi = \frac{v \cdot L}{\upsilon_L \cdot (1 - \psi)}; \quad Sc = \frac{\upsilon_L}{D}$$

Fluidized bed with the voidage ψ:

$$Sh_{fb} = (1 + 1.5 \cdot \psi) \cdot Sh_{sp} \tag{3.5.35}$$

MSMPR crystallizer:

$$k_d(L) = \frac{D}{L} \cdot \left(0.80 \cdot \left[\frac{\bar{\varepsilon} \cdot L^4}{\upsilon_L^3} \right]^{0.2} \cdot Sc^{1/3} + 2 \right) \tag{3.5.36}$$

If bulk diffusion determines crystal growth, the data will lie on the dashed diagonal line. Deviations from this line, especially at small supersaturations, can be due to the increasing influence of the surface integration step on crystal growth rate.

It can be summarized from Figure 3.5.12 (see page 130) that under certain conditions it is possible to obtain fairly useful and reliable growth rates for large crystals in a single crystal cell and in a small fluidized bed for the design of large scale crystallizers.

3.5.5 Deviations from the idealized MSMPR model

The models described above to evaluate the results from MSMPR experiments are only valid if the assumptions defined in Section 3.5.1 hold. Deviations from this idealized MSMPR model may be caused by the following effects:

- kinetics of crystal growth (e.g. size dependent growth, growth rate dispersion)[47,148];
- attrition and breakage caused by mechanical stress of the crystals[78,87,92];
- agglomeration and destruction of agglomerates or dissolution of fines[93];
- classification due to fluid dynamics and product discharge[94];
- insufficient mixing leading to differences of system properties;
- insufficient mixed product removal;
- non steady state operating conditions.

Usually any operating condition will produce some of these effects. For instance increase of stirrer speed causes better mixing but more attrition.

Figure 3.5.13 (see page 132) illustrates qualitatively how some of these deviations from the idealized MSMPR model can influence the steady state crystal size distribution plotted here in the semi-logarithmic population density plot.

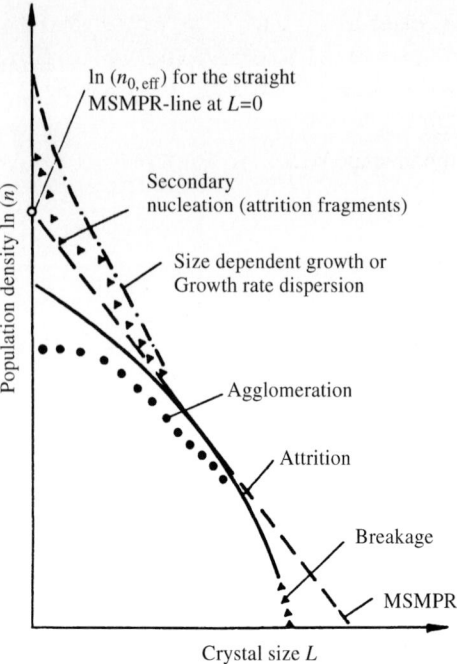

Figure 3.5.13 Deviations from the idealized MSMPR model in a semi logarithmic population density plot

The straight dashed line represents the population density expected according to the idealized MSMPR model. In most cases it is difficult to evaluate the specific reason for the deviation from the linear plot because the same type of deviation can result from several causes. Thus the shape of a population density plot does not on its own explain the particular effect that has caused the deviation in a particular experiment.

(a) Size dependent growth rates

Size dependent growth rate means that the kinetic growth rate of a crystal is a function of crystal size, see section 2.5.

The resulting effect on crystal size distribution is similar to that caused by other effects like growth rate dispersion (where only fast growing crystals appear in the product size distribution), attrition or breakage (since it is more likely that larger crystals collide with the stirrer with a high energy than do smaller ones), or agglomeration or classification.

Therefore in most cases the so-called 'size dependent growth rates' should not be used as a model which is based on a fundamental physical mechanism, but as a simple means to describe the crystal size distribution with its deviation from the straight line.

It is possible to model size dependent growth rates using the ASL-model[95]:

$$G(L) = G_0 \left[1 + \frac{L}{G_0 \tau} \right]^b \qquad (3.5.37)$$

G_0 is the growth rate of very small crystals ($L \to 0$) and the exponent b describes the effects of the crystal size on growth rate and hence size distribution. (It is important to mention that this growth rate G_0 is not suitable for calculating median crystal size L_{50} according to: $L_{50} = 3.67 \cdot G \cdot \tau$).

Equation (3.5.38) is the solution of Equation (3.5.1) using Equation (3.5.37) for the size dependent growth rate:

$$n(L) = n_0 \left[1 + \frac{L}{G_0 \tau} \right]^{-b} \cdot \exp \left[\frac{1 - [1 + (L/G_0 \tau)]^{(1-b)}}{1 - b} \right] \qquad (3.5.38)$$

Figure 3.5.14 (see page 134) illustrates population density plots with 'size dependent' growth rates for various values of the parameter b. For $b = 0$ a straight line is obtained, the same solution as given by the idealized MSMPR model. Negative values for b represents lower growth rates for larger crystals, while the value of b is positive if the growth rates increase for larger crystals.

Mechanical stress
The crystalline material is usually exposed to mechanical stress in any type of crystallizer, caused either by fluid shear or by crystal contact with other solid surfaces such as other crystals, the stirrer, the impeller of a pump, the wall or other parts of the crystallizer.

Possible effects of mechanical stress are:

- attrition at the crystal surface;
- breakage of crystals or agglomerates;
- the possibility that preordered species or surface nuclei may be removed from the crystal surface by shear forces[96].

Mechanical stress can contribute to secondary nucleation and may also influence the growth rate of large crystals.

For laboratory scale crystallizers, attrition effects rather than breakage can be expected to dominate because the stirrer tip speed is usually in the range up to about $4\,\mathrm{m\,s^{-1}}$. At such low tip speeds true breakage does not normally occur

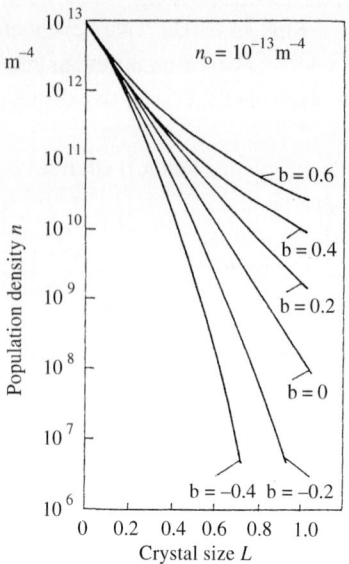

Figure 3.5.14 Semi logarithmic population density plot with size dependent growth rates according to the ASL model[95]

if the clearance between the wall and rotor is large enough. Breakage may start at a collision velocity above approximately $w_{col} = 10\,\mathrm{m\,s^{-1}}$ [97].

The influence of attrition on the effective growth rate $G_{eff} = G_{kin} - G_a$ can be modelled with an attrition rate G_a which operates in the opposite sense to the kinetic growth rate G_{kin} [98]. These authors assume that the kinetic growth rate G_{kin} is only a function of the supersaturation ΔC and that the attrition rate G_a only depends on fluid dynamics with the crystal size L and collision velocity as the most important parameters. By this a maximum possible crystal size L_{max} can be defined, which is the crystal size at which the kinetic growth rate G_{kin} and the linear attrition rate G_a are the same. This rate G_a depends on material properties and the collision velocity w_{col} [97].

Then, the population density distribution can be written as

$$n(L) = n_0 \frac{[1 - (L/L_{max})]^{[(L_{max}/2G_{kin}\tau)-1]}}{[1 + (L/L_{max})]^{[(L_{max}/2G_{kin}\tau)+1]}} \qquad (3.5.39)$$

In Figure 3.5.15 (see page 135) experimental population density distributions for KNO_3 are compared with values calculated from Equation (3.5.39).

The enhanced attrition does not only cause an increase in secondary nucleation, but also directly influences the particle size distribution by limiting the maximum size.

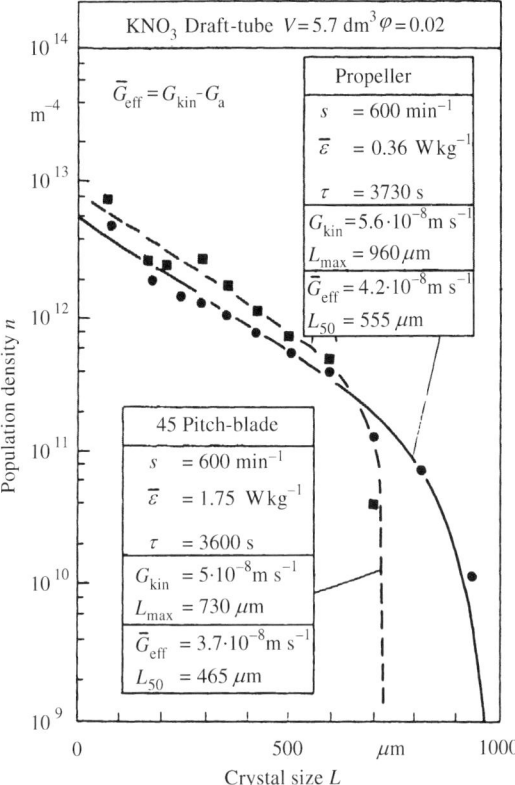

Figure 3.5.15 Influence of attrition on the population density distribution

(b) Fines dissolving, classification

In MSMPR experiments carried out to obtain kinetic information, effects like fines dissolving and classification are, contrary to industrial crystallizers, usually unwelcome because it is difficult to evaluate such experiments as the analysis using the idealized MSMPR model cannot be applied.

In Figure 3.5.16 (see page 137) the effects of fines dissolving and classification of large crystals on the particle size distribution are illustrated[99]. Significant fines dissolution is not usually expected in laboratory MSMPR crystallizers, but significant classification caused by the fluid dynamics of the crystallizer or, more usually, by the product discharge is very common.

In agitated vessels which are not fitted with draft tubes and profiled bottoms, classification always takes place because of the varying solid concentration in the vessel. In this case a representative product discharge is usually impossible

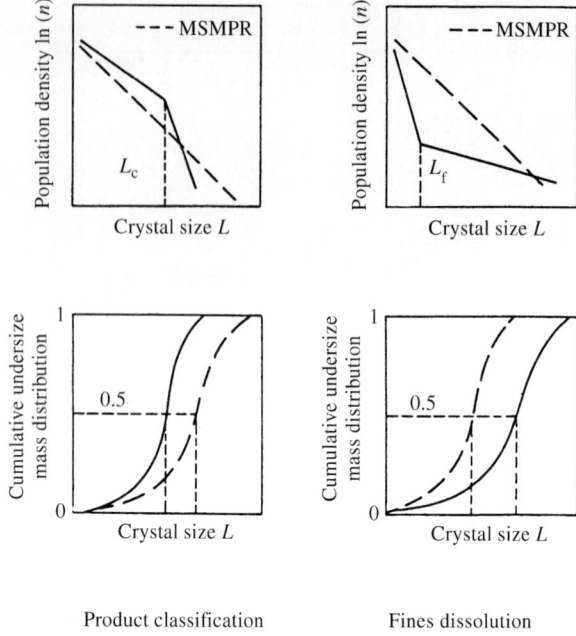

Figure 3.5.16 Effect of fines dissolution and classification of large crystals on particle size distribution[99]

to obtain especially if the density difference between the solid and the liquid phase is high and the crystals are large.

In a DTB crystallizer, classification caused by product removal is also possible if the position of the suction tube or the off take velocity in the tube is not correct. In this case the effects of such classifications on crystal size distribution are not as dominant as they are in a 'normal' agitated vessel. In Figure 3.5.17 (see page 137) the influence of significant classification on the population density distribution resulting from changes in stirrer speed is shown.

If classification does occur it can be characterized by examination of the solid and the liquid phase residence time distributions. The residence time distribution of the solid phase can be determined by 'wash-out' experiments.

Wash-out experiments in a system that is crystallizing are impossible because of crystal growth and attrition. Therefore experiments with model substances which have nearly the same parameters as the investigated substance, such as shape of the particles, size distribution, density difference between solid and liquid, solid concentration and viscosity, are preferable.

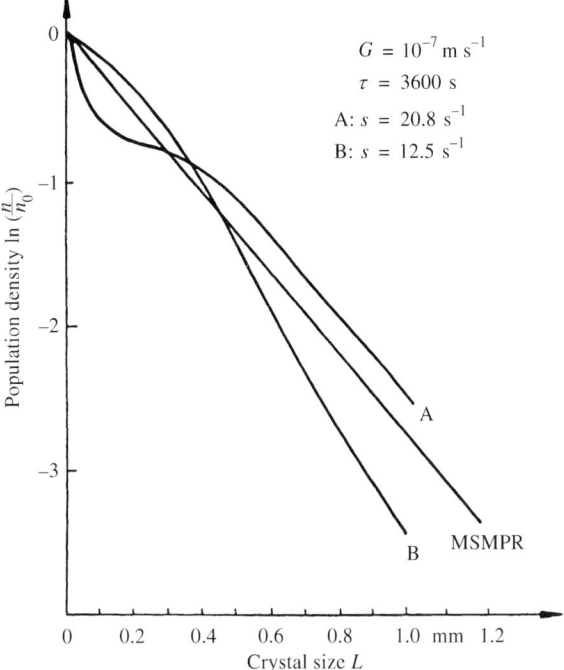

Figure 3.5.17   Effect of wide range classification on population density distribution

Alternatively experiments can be made using the solid crystals in a saturated solution.

The operating conditions of the wash-out experiments, for example, stirrer speed, feed flow rate, evaporation rate, filling height of the crystallizer and discharge properties, must be the same as in the crystallization experiments. The particle size distribution in the crystallizer while the particles are being washed out, with no additional particles being added in the feed stream, is measured as a function of time. The residence times of different particle sizes can then be estimated with Equation (3.5.40) or by a semi-logarithmic plot of the particle mass in every particle class $M(L,t)$ against the time t.

$$\frac{M(L,t)}{M_\alpha(L)} = \exp\left(-\frac{t}{\tau(L)}\right) \tag{3.5.40}$$

with $M_\alpha(L)$ = particle mass at t = 0.

The slope of the lines in the semi-logarithmic plot gives the residence time $\tau(L)$ of each particle size class.

The experimental time depends on the residence time because the solid concentration is reduced during the experiment. With ideal conditions the solid concentration is reduced to 36.8% of the initial particle concentration after one residence time and to about 10% after 2.3 residence times.

3.6 Mechanical material properties relevant for attrition

The determination of nucleation and growth rates from MSMPR-crystallizers for crystals $L > 1 \, \mu m$ is a widely adopted technique. However, when the process of attrition becomes important in suspension crystallizers, most of the simplifying assumptions are violated and it becomes necessary to start with the complete population balance.

Nevertheless, rate equations exist that enable a description of crystallization processes governed by the formation and growth of attrition fragments[82,100–102]. One fundamental requirement for a quantitative description of attrition processes however, is that the relevant mechanical material properties such as the hardness, the shear modulus and the fracture resistance are known.

Measurement techniques for the determination of mechanical properties of various types of solids are well presented in the literature[103]. It is not the aim of this Section to present such established techniques. The objective is to present and discuss procedures for estimating and measuring mechanical material properties which can then be used for estimating attrition rates in suspension crystallizers. The procedures account for the following aspects:

- In experimental procedures substances crystallized from solutions are difficult to handle when they cannot be grown to large sizes ($L > 1$ cm) and when they behave in a brittle manner. Crystals are always anisotropic and it will not generally be possible to grow samples to a particular shape. These aspects impose strong restrictions on measurement techniques. For example, the accurate measurement of the fracture resistance requires that notched bars of a particular shape can be prepared. For this reason, only measurement techniques will be considered that can be performed on crystals several millimeters in size and of arbitrary shape, but with well developed crystal faces.

- For the prediction of size distributions from crystallization processes, a population balance is required. Due to the complex interaction of several rate processes in crystallizers, the population balance is generally written in a one-dimensional form. This means that the rate processes (such as growth rate and agglomeration rate) are directly related to particle sizes which are only one-dimensional. In terms of its material properties, the crystal is therefore considered to be isotropic. For the description of the relevant material properties for attrition, it will frequently be required that the anisotropic behaviour of crystals can be described by average quasi-isotropic properties. This also implies that other material characteristics affecting the attrition rate (flaw size distribution, particle shape) must be considered constant for a given size class in the population balance.

The following considers those mechanical material properties which are considered to be mainly responsible for the attrition behaviour of brittle solids[104]. These are measures of the ability to deform elastically, plastically and a measure of its resistance to form cracks.

3.6.1 Elasticity

The most widely-used method for measuring the elastic constants is ultrasonic wave transmission[103]. An extensive collection of elastic constants of organic and inorganic crystals is found in Landolt-Börnstein[105], where a summary is given of all measurement techniques used so far.

An isotropic material has two independent elastic constants, the shear modulus μ and Poisson's ratio v, from which all other possible constants can be calculated. Anisotropic crystals may have between 3 (in the cubic system) and 21 (in the triclinic system) elastic constants, depending on their symmetry properties. From their values, average, quasi-isotropic constants can be calculated according to Voigt, Reuss and Hill[106]. We here focus on a few fundamentals which are important for attrition.
Poisson's ratio v is given by:

$$v = \frac{E}{2\mu} - 1 \tag{3.6.1}$$

with E as Young's modulus.

3.6.2 Plastic flow and fracture

The following Vickers indentation measurement is used for the estimation of the plastic flow and fracture behaviour of brittle solids[107–110].

Experimental

For the experimental procedure, particles with well developed crystal faces and sizes larger than 1mm are required. The substances discussed in the following section were obtained by adding single crystals of approximately 100 μm into a glass beaker, where they were allowed to grow by cooling the aqueous solutions at low rates (<1 K/h). More sophisticated methods are required when the substances do not grow to the desired size or develop porous structures with badly developed faces. For the identification of the crystal faces it is desirable that crystals grow to sizes of approximately 5 mm. To avoid further crystal-lization once removed from the beaker, inorganic crystals can be introduced into a non-solvent (e.g. n-hexane), which in turn can be removed by warm air. Frequently, however, it will be sufficient to remove the solution with absorbent paper. The crystals should be handled with care, since the following experi-mental procedure requires areas of approximately 1000 μm^2 on the crystal face which is smooth on a microscopic scale.

The crystals are fixed on a glass plate with a hardening glue. This glass plate is then fixed in a device (Figure 3.6.1a), which enables the vertical alignment of a given crystallographic face under a macroscope. The crystal faces can be identified under the macroscope with the help of a goniometric eyepiece.

Vickers-indentation experiments are then performed on the aligned crystal face (Figure 3.6.1b). A micro-hardness measurement device is required to determine the relevant material properties, which should cover a load range from 10^{-3} to 1 N. A typical experiment consists of stepping the indenter into the specimen within 5 seconds, after which the maximum load is applied for 15 seconds. The indenter is then withdrawn and the permanent impression is viewed by optical microscopy. Cracks that may form around the indentation can be detected once they reach a size of approximately 1 μm. A mirror mounted in the alignment device (Figure 3.6.1) enables good illumination of transparent crystals from below.

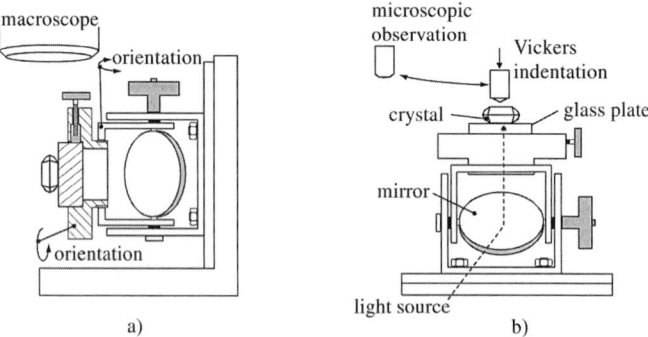

Figure 3.6.1   Alignment of crystals for indentation experiments

The hardness is obtained by dividing the indentation load by the area of indentation formed. Hardness measurements should generally be made when no cracks are visible around the indentation. The Vickers hardness is calculated from

$$H_V = 1.854 \frac{F}{d_1 d_2} \qquad (3.6.2)$$

where F is the indentation force and d_1 and d_2 the two diagonals of the plastic indentation. The factor 1.854 results from the geometry of the indenter, which is a pyramid having an included angle of $136°$.

Anisotropy of indentations and crack patterns

The experimental observations summarized in this section are based on a comparison of ten different organic and inorganic substances where indentation tests have been performed on a total of 27 different crystallographic faces (see Table 3.6.1 on page 142). By comparing the hardness as well as the crack patterns formed by indentations, it is possible to group the material behaviour of crystalline substance into three classes[110].

1) Brittle crystals

Covalently-bonded crystals are generally hard and brittle. For these crystals the following results are typically obtained from an indentation experiment:

- brittle crystals have a low ratio of E/H_V and generally a high value of H_V;
- the crack patterns observed on these crystals are radial as well as lateral (see Figure 3.6.2 on page 142);
- the crack patterns are very similar on different faces.

From the ten substances investigated, five can be attributed to this group: citric acid, magnesium sulphate, potash alum, potassium sulphate and tartaric acid. They have a relatively high hardness (H_V between 454×10^6 and 1502×10^6 Pa) and a ratio E/H_V between 25 and 36.

Typical crack patterns on these crystals are shown in Figure 3.6.3 (see page 143) for the $\{1\,1\,1\}$ face of potash alum and the $\{1\,0\,1\}$ face of citric acid.

2) Semi-brittle crystals

Semi-brittle crystals are neither typical ionic nor covalent crystals. As a consequence, larger values of E/H_V are generally observed for these crystals. Three substances which can be attributed to this class are ammonium sulphate, potassium nitrate and thiourea. They have a hardness H_V between 137×10^6 and 355×10^6 Pa and a ratio of E/H_V between 59 and 71.

Table 3.6.1 Substances investigated[110]

	Formula	Examined forces
Ammonium sulphate	$(NH_4)_2SO_4$	$\{0\,0\,1\}$, $\{0\,1\,0\}$, $\{1\,1\,0\}$
Citric acid	COH–COOH $(CH_2COOH)_2 \cdot H_2O$	$\{0\,1\,1\}$, $\{1\,0\,1\}$, $\{1\,1\,0\}$
L (+) tartaric acid	$(CHOH)_2(COOH)_2$	$\{1\,0\,0\}$, $\{1\,1\,0\}$
Magnesium sulphate	$MgSO_4 \cdot 7H_2O$	$\{0\,1\,0\}$, $\{1\,1\,0\}$, $\{1\,1\,1\}$
Potash alum	$KAl(SO_4)_2 \cdot 12H_2O$	$\{1\,0\,0\}$, $\{1\,1\,0\}$, $\{1\,1\,1\}$
Potassium chloride	KCl	$\{1\,0\,0\}$
Potassium nitrate	KNO_3	$\{0\,1\,0\}$, $\{0\,1\,1\}$, $\{0\,12\}$, $\{1\,1\,0\}$
Potassium sulphate	K_2SO_4	$\{0\,1\,0\}$, $\{0\,21\}$, $\{1\,1\,0\}$, $\{1\,11\}$
Sodium chloride	NaCl	$\{1\,0\,0\}$
Thiourea	$CS(NH)_2$	$\{0\,0\,1\}$, $\{0\,1\,0\}$, $\{1\,1\,0\}$

When one considers the indentation as well as the types of cracks formed, a strong anisotropic behaviour can be observed for these crystals. Typical indentations and crack patterns on a potassium nitrate crystal are drawn in Figure 3.6.4 (see page 143). The indentation diagonals are very different on the $\{1\,1\,0\}$ and $\{0\,1\,0\}$ faces. Because of the anisotropic plastic flow, lines corresponding to regions of surface pile-up can be observed. Radial cracks are formed on these faces, regardless of the orientation of the Vickers-diagonals. The faces $\{0\,1\,1\}$ and $\{0\,2\,1\}$ of potassium nitrate show a different behaviour. Only lateral cracks form on these faces; this is understandable,

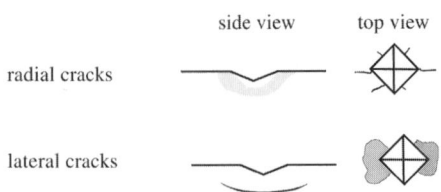

Figure 3.6.2 Schematic of the observed radial and lateral cracks

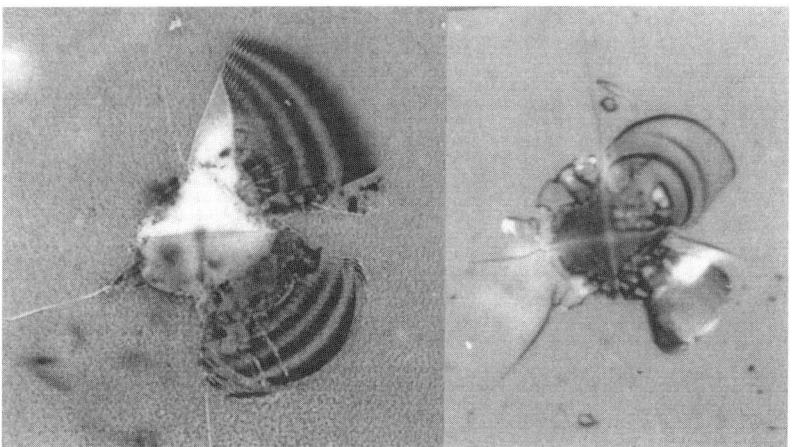

Figure 3.6.3 Indentations on the {1 1 1} face of potash alum (left) and the {1 0 1} face of citric acid (right) at a load of 0.1 N

since these cracks run roughly along the cleavage plane {0 1 1}[111]. It is important to note that cleavage is most frequently observed in semi-brittle materials[112]. On brittle substances this tendency is generally much less pronounced, since their (covalent) bonds are highly directional and hinder the movement of dislocations and consequently plastic flow. Although the indentations on semi-brittle substances may be strongly anisotropic, strong variations of the hardness values on different crystallographic faces (by more than 20%) cannot be observed.

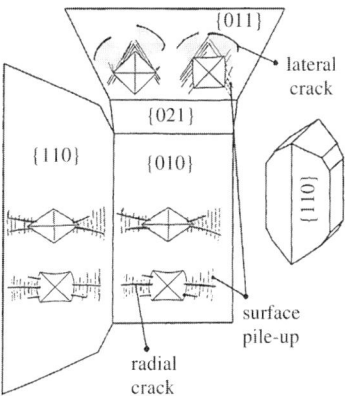

Figure 3.6.4 Illustration of crack patterns and surface pile-up for indentations on potassium nitrate

3) Ductile crystals

Ionic bonds are non-directional and this is the reason why dislocations can glide easily, resulting in low hardness values. The alkali halides sodium chloride and potassium chloride have a hardness H_V of 166×10^6 and 91×10^6 Pa and a ratio of E/H_V of 222 and 265 respectively. No cracks can be formed on these crystals by a Vickers indentation test, even at loads of 30 N. A Vickers indentation test should, however, not lead to the misleading conclusion that these substances will not form attrition fragments under impact conditions.

The ability of ionic crystals to deform plastically depends strongly on the strain rate. In a conventional Vickers indentation experiment the load is applied slowly in a quasi-static manner. Under impact conditions in a crystallizer, however, the strain rate is significantly higher and dislocations have less time to move, resulting in a smaller ability to deform plastically[113]. A dynamic measurement will therefore result in significantly higher values of the hardness for ionic crystals (up to a factor of five[114]). For these substances, a quasi-static indentation test is insufficient to determine the relevant mechanical material properties that determine an attrition process.

From the above observation it can be expected that it is possible to extract the following information from indentation experiments:

- An indentation test can determine whether a substance behaves in a brittle, semi-brittle, or ductile manner under quasi-static conditions. The quantitative measure of this behaviour is the ratio of Young's modulus and hardness, the qualitative measure is the characteristic crack pattern.
- Indentation fracture is a helpful experimental technique for the observation of anisotropic material behaviour.

Estimation of the fracture surface energy

Section 3.5.3 has shown that besides the hardness and the shear modulus, the fracture resistance is also an important parameter for attrition phenomena. The following procedure enables a more quantitative analysis of this property and the crack patterns presented in the previous section.

The work performed on a specimen during loading of the Vickers indenter can be estimated from[110]:

$$W = 6.0 \cdot 10^{-2} \sqrt{\frac{F^3}{H_V}} \qquad (3.6.3)$$

After the Vickers test has been performed, cracks may become visible around the indentation.

There is no explicit solution available for the stress field of an elastic-plastic indentation (see the review by Tabor[115]). However, the stress distribution in a linear elastic material loaded by a point force (Boussinesq's stress field[116]) can explain the fracture phenomena frequently observed in indentation experiments[104]. Assuming that the indenter acts as a point force, the two important aspects of the stress field are that large tensile stresses as well as the maximum shear stresses are generated directly below the indenter (Figure 3.6.5a). As a result of these stresses, one would expect radial (caused by the tensile components) and lateral (caused by the shear components) cracks to develop from an existing flaw just below the indenter (Figure 3.6.5b).

In solution grown crystals, flaws will generally exist on the molecular scale (dislocations) as well as on the microscopic scale (inclusions); the larger the flaw, the lower the stress required for crack growth. The development of cracks will therefore be a function of the flaw size distribution which is an important parameter for the fracture resistance.

One concept is to estimate the fracture toughness from indentation experiments[109,117]. However, these estimations require an analysis of the stress distribution at the crack tip, which is difficult to obtain.

Another concept is to relate the work of indentation to the newly created surface, A, formed by cracking[102,110]. This analysis is based on the assumption that the newly created surface formed by the cracks is proportional to the elastic strain energy $W_{elastic}$ stored in the specimen (Rittinger's law):

$$W_{elastic} = A \frac{\Gamma}{K_r} \tag{3.6.4}$$

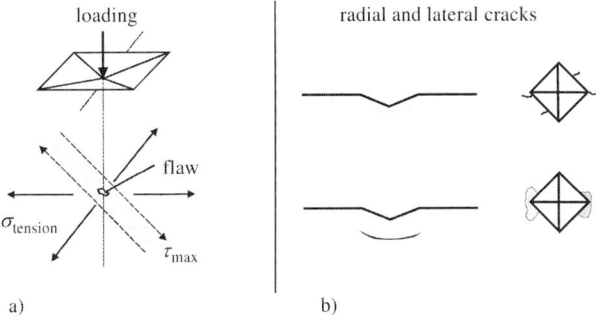

Figure 3.6.5 Growth of cracks: a) initiation of crack growth due to tensile and shear components; b) lateral and radial cracks

The material property which determines the proportionality constant is the fracture surface energy Γ. The efficiency constant K_r takes into account that only tensile or shear stresses can cause fracture. This results in radial and lateral cracks which depend on the work of indentation, W^{110}. Cracks become visible when the work of indentation exceeds a critical value (Figure 3.6.5b).

This critical work, W_C, can be related to the fracture surface energy by assuming that the area where cracks have formed is approximately the same size as the area of the indentation[118].

$$\frac{\Gamma}{K_r} \approx \frac{1}{10} \frac{W_C^{1/3} H_V^{5/3}}{\mu} \qquad (3.6.5)$$

From this crude estimation, which disregards the types of cracks formed, the ratio Γ/K_r is obtained representing the fracture surface created divided by the total elastic strain energy in the specimen.

The critical work, W_C, where cracks become visible around the indentation is determined by performing a series of indentation experiments at different loads on a given crystallographic face. It will generally be sufficient to perform ten indentations at a given load on one crystallographic face. In Figure 3.6.6 (see page 147) the fraction of indentations where cracks can be observed is plotted as a function of the applied load for different faces of thiourea. From these plots, the force where around 50% of the indentations cracks become visible, $F_{0.5}$ can be obtained. This is taken to be the characteristic value for the calculation of the critical work W_C:

$$W_C = 6.0 \cdot 10^{-2} \sqrt{\frac{F_{0.5}^3}{H_V}} \qquad (3.6.3a)$$

A quasi-isotropic value of the critical work is obtained by calculating the geometric average from the individual values of each face.

It can be seen from Figure 3.6.6 that the fraction of indentations where cracks become visible depends on the face and increases sharply with increasing load. This increase can be well reproduced on the same face of different specimens. From these observations it can be concluded that the procedure used here is not very sensitive to the flaw size distribution in the specimen. The reason for these observations is that visible cracks are larger (at least several microns) than the flaws from which they nucleate.

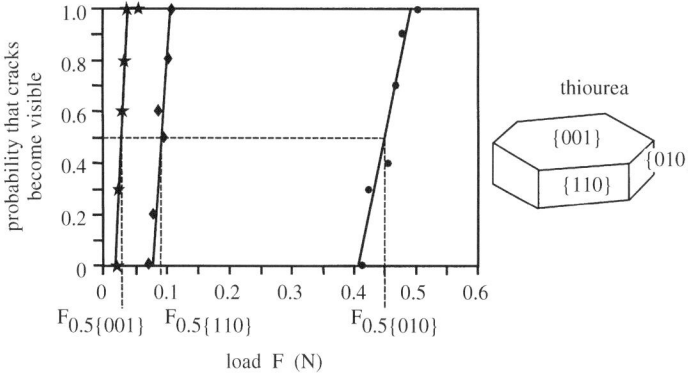

Figure 3.6.6 Determination of the probability that cracks become visible around indentations of thiourea

3.6.3 Problems in the determination of the material properties relevant for attrition

From an experimental as well as from a theoretical point of view, the mechanical properties of brittle materials can more easily be characterized than semi-brittle or ductile substances. The main reason is that the ability to deform plastically will be more strain rate dependent for the latter materials. Whether a substance behaves in a brittle manner or not will depend on the ratio of Young's modulus and hardness. For low ratios of E/H_V, the substance is brittle. For this case it can be shown[118] that the fracture surface energy can be estimated by assuming that it is proportional to the surface energy. The estimation of the surface energy, however, only considers elastic behaviour[119]. The suggested equation for the fracture surface energy is:

$$\frac{\Gamma}{K_r} \approx 1.7\,E\,l_0 \approx 1.7\,E \cdot \sqrt[3]{\frac{M}{\rho_C N_A n}} \tag{3.6.6}$$

where the average distance of elements, l_0, in the crystal can be estimated from n as the number of elements. N_A is the Avogadro number.

Table 3.6.2 (see page 148) lists the material properties that have been estimated and measured from the procedures given above. The substances are given in order of increasing ratio of E/H_V. It can be seen that the values of Γ/K_r obtained from the two different estimations of the fracture surface energy do not deviate by more than a factor of two for the brittle substances ($E/H_V < 40$). Considering the crude approximations made in the analysis, a better correlation cannot be expected. It will be required especially for semi-

147

Table 3.6.2 Material properties of brittle, semi-brittle and ductile crystals

Material properties Substance	H_V $*10^{-6}$ [N m^{-2}]	μ $*10^{-9}$ [N m^{-2}]	E $*10^{-9}$ [N m^{-2}]	v [–]	E/H_V [–]	W_C $*10^{10}$ [J]
(L +)-tartaric acid	1030	9.58	25.4	0.32	25	36
Potash alum	754	7.96	20.3	0.27	27	7
Citric acid	454	4.71	12.6	0.35	28	35
Potassium sulphate	1502	17.4	44.1	0.27	29	30
Magnesium sulphate	649	9.06	23.6	0.31	36	48
Thiourea	137	3.08	8.09	0.33	59	1800
Ammonium sulphate	355	8.90	23.4	0.32	66	41
Potassium nitrate	265	7.17	18.9	0.32	71	59
Sodium chloride	166	14.7	36.9	0.25	222	–
Potassium chloride	91	9.44	24.1	0.28	265	–

Material properties Substance	ρ_C [kg m^{-3}]	$\tilde{M}$ [g mol^{-1}]	n [-]	I_o*10^{10} [m]	Γ/K_r (Equation 3.6.6) [J m^{-2}]	Γ/K_r (Equation 3.6.5) [J m^{-2}]
(L +)-tartaric acid	1759	150.1	16	2.1	8.95	16.8
Potash alum	1760	474.4	48	2.1	7.26	7.0
Citric acid	1542	210.1	24	2.1	4.54	8.7
Potassium sulphate	2662	174.3	7	2.5	18.7	16.4
Magnesium sulphate	1680	246.5	27	2.1	8.37	9.1
Thiourea	1405	76.12	8	2.2	3.08	6.7
Ammonium sulphate	1769	132.1	15	2.0	8.06	3.2
Potassium nitrate	2109	101.1	5	2.5	8.09	2.8
Sodium chloride	2163	58.44	2	2.8	–	–
Potassium chloride	1984	74.56	2	3.1	–	–

brittle and ductile materials therefore, that measurement techniques are used which enable the estimation of mechanical properties under impact conditions.

Although the measurement technique presented here aims to predict attrition rates in suspension crystallizers, the important aspect of the influence of properties of the surrounding liquid (especially supersaturation) has not been considered.

From a thermodynamic point of view it is necessary to distinguish between the generation rate of attrition fragments and the number of fragments which can grow in supersaturated solutions. The process of fracture always results in a plastically deformed region close to the fractured surface. Even in highly brittle materials such as quartz, plastic deformation takes place, resulting in a temperature increase at the crack tip of 4000 K[120]. The non-equilibrium at the crystal surface will result in the dissolution and rearrangement of the plastically deformed region, or in the complete dissolution of the attrition fragment[100,101,118,121]. The generation rate of attrition fragments is therefore expected to depend on the supersaturation in the solution. So far, however, no measurement techniques have been developed which could distinguish between the amount of particles produced and their growth behaviour.

Methods of nucleation rate measurement

4

This chapter will show that the measurement of rates of activated nucleation (in the absence of any mechanical effect and controlled by an activation energy) requires different experimental equipment and techniques in comparison to crystal growth.

4.1 Introduction

It is difficult to define and measure nucleation rates because the term 'nucleus' is used for a variety of solid species present in a solution. As a rule the nucleation rate B is defined as the total number of particles N_{tot} generated in a certain volume ΔV of constant supersaturation ΔC and in a certain time Δt during which ΔC should remain constant:

$$B = \frac{N_{tot}}{\Delta V \cdot \Delta t} \tag{4.1.1}$$

or according to the MSMPR modelling

$$B_{\lim L \to 0} = \frac{N_{tot}}{\Delta V \cdot \Delta t} \equiv n_0 \cdot G \tag{4.1.2}$$

New particles are born as attrition fragments at low supersaturation ($\sigma < \approx 0.1$) or 'nuclei' by activated nucleation at high supersaturation. In Figure 4.1.1 (see page 152), the median crystal size L_{50} is plotted against the relative supersaturation. Experimentally-determined median sizes of crystals produced in laboratory or industrial crystallizers (only in a one-stage process without classification and fines destruction) can be found in the shaded area between several mm and several nm[96,97,122]. The experimentally-determined relative supersaturations σ are in the range $10^{-4} < \sigma < 10^6$. Crystals larger than approximately $100\,\mu m$ are reduced in size by attrition induced by particle collisions. Only large attrition fragments are 'effective' nuclei, see Section 3.5.3. It has previously been shown that in industrial crystallizers

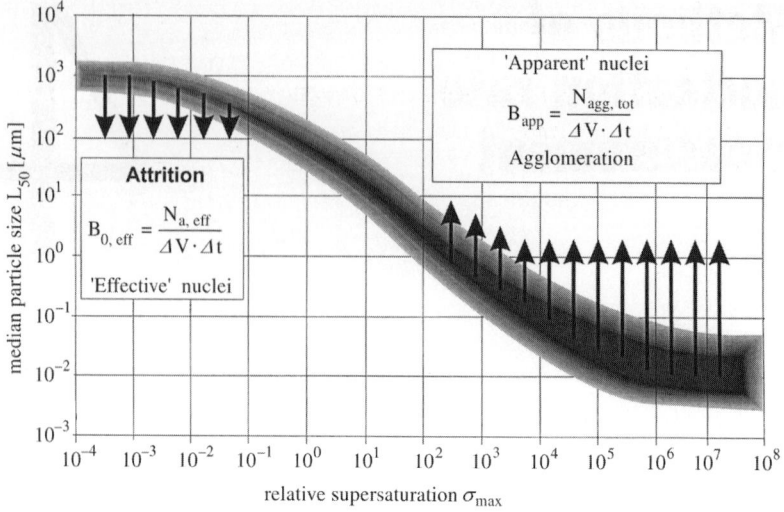

Figure 4.1.1 Experimentally determined median crystal size for various materials (shaded area[28,123,150]) and different nucleation mechanisms

attrition fragments are generated in the size range between a few micrometers up to approximately 100 μm and that only a small percentage (roughly 1%) of them (large fast growers) grow into the product size range. Small attrition fragments are 'not effective' nuclei. The effective rate of secondary nucleation, $B_{0,eff}$, is proportional to the collision frequency f_c, the total number $N_{a,tot}$ of the attrition fragments, and the ratio ($N_{a,eff}$ /$N_{a,tot}$) which depends on supersaturation (see Section 3.5.3):

$$B_{0,eff} \sim f_c \cdot N_{a,tot} \left(\frac{N_{a,eff}}{N_{a,tot}} \right)$$ (4.1.3)

In the case of $\sigma > 0.1$ surface nucleation on crystals present in the solution can take place, and at $\sigma \gg 0.1$ many nuclei are born by primary (homogeneous and/or heterogeneous) nucleation. The rates of these kinds of activated nucleation depend strongly on the supersaturation $S = 1 + \sigma$ according to:

$$B_{act} = A \cdot \exp\left(-\frac{B}{(v \ln S)^a} \right)$$

with $a = 2$ for primary nucleation and $a = 1$ for the surface (or activated secondary) nucleation[96].

In the case of high supersaturation ($\sigma > 100$) many nuclei are generated which leads to very high particle concentrations $N_{tot}/\Delta V$ with the result that a massive and fast agglomeration can take place which may strongly reduce the number N_{tot} of particles. Such 'nuclei' are agglomerates composed of a certain number of primary particles in the nanometer range. When the particle concentration is sufficiently reduced and consequently the agglomeration rate has become small $((dN/dt)_{agg} \sim N^2)$, the number of agglomerates $N_{agg,tot}$ generated per unit volume and time can be termed an apparent nucleation rate B_{app}:

$$B_{app} = \frac{N_{agg,tot}}{\Delta V \cdot \Delta t} \qquad (4.1.4)$$

However, the supersaturation σ is only constant in a very small volume element $\Delta V \to 0$ during a very short duration $\Delta t \to 0$. As a rule the supersaturation differs from point to point at a high σ-level, depending on micro-mixing, and decreases rapidly with time (see Figure 4.1.2) due to crystallization kinetics, both nucleation and growth. The time of nucleation is only defined for the ideal relationship $\sigma = f(t)$, which is characterized by a sudden increase and a rapid decrease of supersaturation. Since a certain amount of time for mixing and heat and mass transfer is always required to achieve different levels of σ, the nucleation time Δt is badly defined for the real curve $\sigma = f(t)$.

These considerations show the general difficulty in defining and measuring nucleation rates and explain many contradictory results published in the literature. They also make clear that strictly speaking it is not correct to omit the birth function $B(L)$ and the death function $D(L)$ in the population balance because strong agglomeration at high supersaturation ($\sigma > 100$) and severe attrition of large crystals with maximum size of attrition fragments up to $100\ \mu m$ (depending on fluid dynamics) take place for $\sigma < 0.1$. Therefore, we cannot expect a straight line in a semi logarithmic ln n, L-diagram in the range

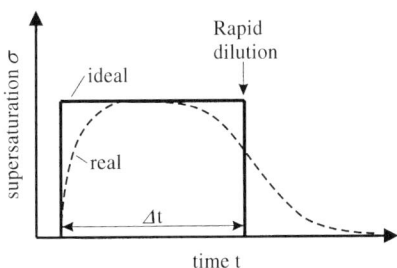

Figure 4.1.2 Comparison of an ideal and a real course of the supersaturation during a nucleation experiment

$0 < L < \infty$, and the often used nucleation rate $B = B_{app}$ for $\sigma > 100$ or $B = B_{0,eff}$ for $\sigma < 0.1$ with:

$$B = n_0 \cdot G \qquad (4.1.5)$$

is not a real nucleation rate but only a helpful magnitude (the 'effective' or 'apparent' nucleation rate, see later) in the equation:

$$L_{50} = 3.67 \cdot \sqrt[4]{\frac{G}{6\alpha}\left(\frac{\varphi}{B}\right)} \qquad (4.1.6)$$

based on the MSMPR-modelling.

For chemical engineering purposes, this modelling is often sufficiently accurate if the product crystals are in the size range between say 100 μm and $\approx 1000\,\mu m$ and the largest attrition fragments are smaller than 100 μm. Then $B_{0,eff}$ is an 'effective' nucleation rate.

Equation (4.1.6) is not valid for particles generated at high supersaturation in the nanometer range because the strong agglomeration is not taken into account. However, if the 'apparent' nucleation rate, $N_{agg,tot}/(\Delta V\,\Delta t)$, after the agglomeration process is taken as B_{app} the MSMPR-model may give reasonable results for products in the size range between 1 μm and 100 μm because agglomeration rates are small and attrition is negligible with respect to the small size of the product crystals.

Since attrition controlled secondary nucleation has already been described in previous sections, we will now focus on activated nucleation. The laws of activated (e.g. homogeneous, heterogeneous and crystal surface induced) nucleation and of perikinetic agglomeration in combination with the already described phenomena of attrition allow rough estimates of the dominant mechanisms that determine the number of particles present in a crystallizer. If the supersaturation is constant in a certain volume element or volume during the time τ the number N of nuclei per unit volume is given by:

$$N = B \cdot \tau \qquad (4.1.7)$$

Let us assume that every collision of two particles of the same size leads to an agglomeration event in order to make clear the importance of agglomeration (this requires the absence of any repulsive forces between the particles). A combination of this equation with the Smoluchowski law of perikinetic agglomeration[124] for a monosized suspension with the particle size L according to:

$$\left(\frac{dN}{dt}\right)_{col} = -4\pi \cdot D_{AB} \cdot L \cdot N^2 = -4\pi \cdot D_{AB} \cdot L \cdot \tau^2 \cdot B^2 \qquad (4.1.8)$$

154

leads with $B = -(dN/dt)_{col}$ to the critical nucleation rate:

$$B_c = \frac{1}{4\pi \cdot D_{AB} \cdot L \cdot \tau^2} \tag{4.1.9}$$

Real nuclei are only measured if:

$$\left(\frac{dN}{dt}\right)_{col} \ll B_c \tag{4.1.10}$$

In the case:

$$\left(\frac{dN}{dt}\right)_{col} \geq B_c \tag{4.1.11}$$

nuclei can agglomerate with the result that apparent nucleation rates are measured. With a residence time of $\tau = 1$ s and a diffusivity $D_{AB} = 10^{-9}\,m^2\,s^{-1}$, the critical rate $B_{hom,c}$ of primary homogeneous nucleation is:

$$B_{hom,c} = 10^{15}\,\text{nuclei}\,m^{-3}\,s^{-1} \quad \text{for } L = 100\,nm \tag{4.1.12}$$

However, this value increases strongly in the case of shorter residence times and smaller sizes (for example, $B_{hom,c} = 10^{28}\,\text{nuclei}\,m^{-3}\,s^{-1}$ for $\tau = 10^{-6}$ s and $L = 10$ nm).

In Figure 4.1.3 (see page 156) the dimensionless supersaturation $\Delta c/c_c$ is plotted against the dimensionless solubility c^*/c_c. The curve valid for $B_{hom,c} = 10^{15}\,\text{nuclei}\,(m^{-3}\,s^{-1})$ is based on the diffusivity $D_{AB} = 10^{-9}\,m^2\,s^{-1}$, the surface tension[150]:

$$\gamma_{CL} = \frac{K(kT)}{d_m^2}\ln\left(\frac{c_c}{c^*}\right) \quad \text{with } 0.31 < K < 0.414 \tag{4.1.13}$$

and the supersaturation $S = c/c^*$, based on concentrations. It is important to mention that the difference between the supersaturation in concentrations $S = c/c^*$ and the supersaturation in activities $S_a = a/a^* = \gamma c/(\gamma^*c^*)$ can be significant at high concentrations. In the area marked 'attrition controlled', rates of activated nucleation are so small that they can be neglected. Above the curve valid for $B_{hom,c} = 10^{15}\,\text{nuclei}\,(m^{-3}\,s^{-1})$, strong agglomeration can occur. This has been confirmed experimentally for $BaSO_4$ by Schubert[125]. In the area below this curve and above the attrition area, real rates of activated nucleation can be measured.

It should be noted that the factor K in Equation (4.1.13) has a strong influence on the rates of activated nucleation. It has been shown that this factor should be in the range from 0.31 up to 0.414[97].

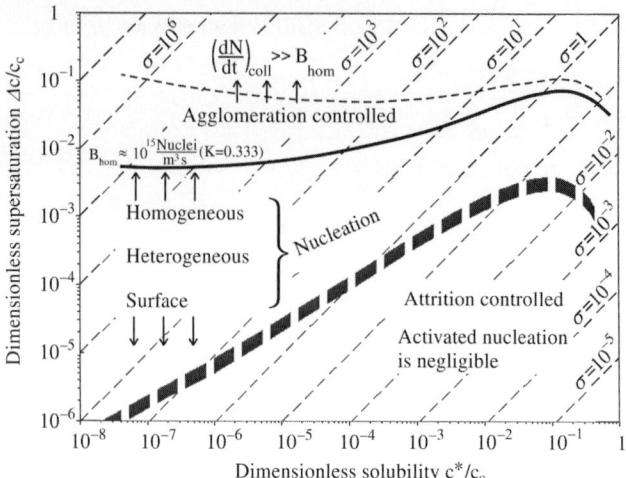

Figure 4.1.3 Approximate areas in which agglomeration, primary nucleation and secondary nucleation are dominant (valid for $D_{AB} = 10^{-9}\,m^2\,s^{-1}$; $v = 1$; $T = 298\,K$; $\gamma_{CL} = 0.33\,kT/d_m^2\,\ln(c_c/c^*)$ and calculated in concentrations)

4.2 Metastable zone width measurement

The induction period of nucleation, t_N, is the period elapsed from attainment of a given supersaturation up to the formation of a critical nucleus. As it is so far impossible to indicate experimentally the formation of critical nuclei, it is necessary to wait until they grow to visible size after time t_G. The total induction period, t_{ind}, consisting of the sum of t_N and t_G, can however be measured experimentally and represents the time that elapses from attainment of super-saturation in the system up to the appearance of the new phase. If $t_N < t_G$ for any supersaturation, then the nucleation step has only a negligible effect on t_{ind} and no data on nucleation can be derived from the dependence of t_{ind} on the supersaturation; the existence of a metastable region cannot be considered for such a system. If, however, $t_N > t_G$, then t_{ind} yields information on nucleation and the metastable zone width measurements become a useful tool in nuclea-tion studies[64].

Metastable zone width measurements are a simple method for the estimation of relative nucleation data, suitable in particular for study of the effect of various parameters on nucleation kinetics. The principle of the measurement is as follows.

The phase diagram of a two-component solid–liquid system with a positive temperature coefficient of solubility, dw*/dT, is given in Figure 4.2.1 (see page 157). Cooling a solution from initial point A leads to point B on the solubility

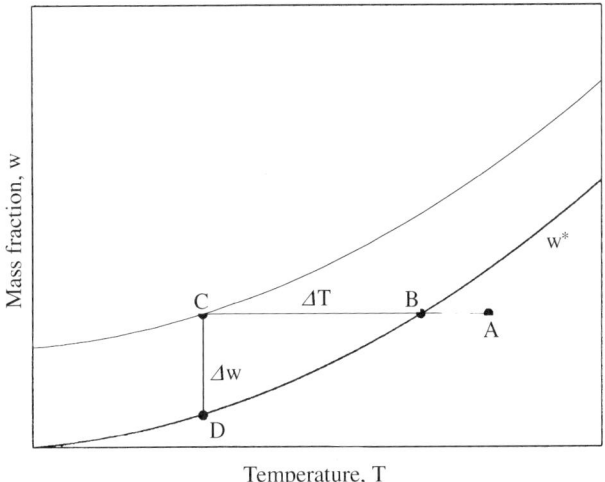

Figure 4.2.1 Principle of the metastable zone width measurement

curve. At this point, the solution is just saturated and in equilibrium with any solid phase that is present. Further cooling results in a state in which the concentration of the solute is greater than the equilibrium concentration at the given temperature and the solution is thus supersaturated. Nevertheless, the solid phase is not precipitated spontaneously in a sufficiently short time until point C is attained, this point corresponding to the metastability boundary of the solution. The position of the metastability boundary is expressed by the maximal attainable undercooling, $\Delta T_{max} = T_B - T_C$, corresponding to the maximum attainable supersaturation, $\Delta w_{max} = w_C - w_D$. These terms are related by the equation:

$$\Delta c_{max} \cong \Delta T_{max} \left(\frac{dw^*}{dT} \right) \tag{4.2.1}$$

Nucleation in solutions of readily soluble substances is controlled by laws described in more detail in Section 4.1. Clusters of solute molecules form and disintegrate as a result of local concentration fluctuations. For each solution supersaturation, a critical cluster size can be determined—the critical nucleus—that is in equilibrium with the surrounding medium and has the same probability of both growth and disintegration. On the other hand, every concentration and temperature of the solution (even in the region of undersaturated solutions) possesses a corresponding steady-state average size of solute clusters. If the state of the solution changes, so also does the state of

157

aggregation of particles. This change occurs, however, at a limited rate so it may be delayed in comparison with the change of state of the system[64,71]. So it is clear that the width of the metastable zone (or induction time necessary for the clusters to reach the critical size) depends on many factors such as temperature, cooling rate, agitation, thermal history of solution, presence of solid particles and of admixtures.

Figure 4.2.2 shows, for example, the most important effect of the cooling rate[64,73]. With slow cooling, the size of clusters in solution can follow the change in the solution state so that the metastable zone width is narrower. On the other hand, with rapid cooling the structure changes in solution are slower than the change of the temperature and the metastable zone width becomes wider.

In this way, the induction period of nucleation and the metastable zone width are interconnected and dependent on the conditions of the experiment. The position of the horizontal line as well as that of the vertical line in Figure 4.2.3 (see page 159) may move in a broad region (but their intersection must follow the curve).

As the metastable zone width depends on many variables, it is necessary to keep the experimental conditions constant within one set of measurements and record all necessary values of variables. In addition to the cooling rate (or, in general, the supersaturation rate), these conditions are:

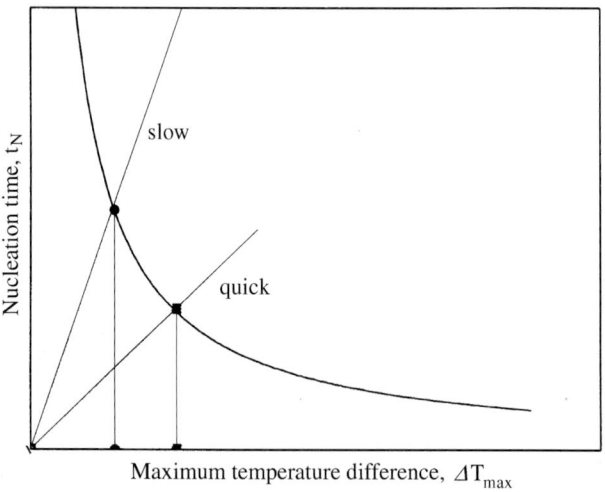

Figure 4.2.2 Dependence of the metastable zone width on cooling rate[73]

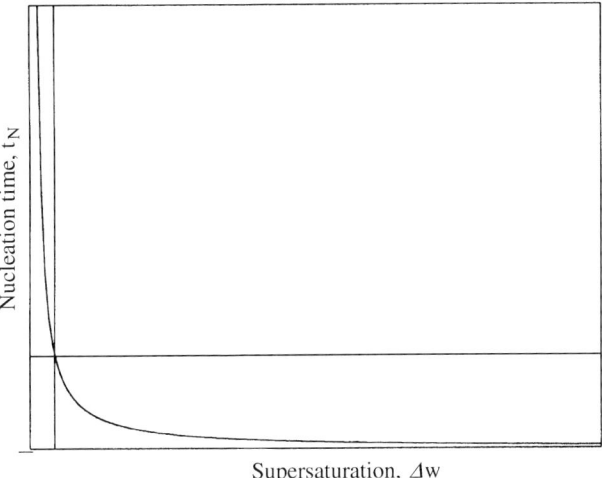

Figure 4.2.3 The dependence of the induction period of nucleation on supersaturation

- The composition of the solution: Even traces of impurities can significantly affect the metastable zone width;
- The temperature: $\Delta T_{max}/T^*$ always decreases with increasing saturation temperature[64];
- The physical purity of the solution: Presence of solid particles (foreign particles of solute crystals) usually leads to a narrower metastable;
- The thermal history of the solution: Solutions that are maintained at a temperature significantly higher than the equilibrium temperature for a sufficiently long time have broader metastable zones than solutions whose temperature has not increased above the saturation temperature[71] (see Figure 4.2.4 on page 160). This effect can reduce the reliability of the metastable zone width measurements, not only when repeating experiments with an old solution (when the first measurement has to be disregarded), but even in MSMPR measurements with recirculating liquor.
- Mechanical action on the solution: Agitation or ultrasonics always lead to narrowing of the metastable zone (see Figure 4.2.5 on page 160).
- Many other factors, such as the volume of solution (nucleation is less probable in small droplets), viscosity (high viscosity tends to prevent nucleation), quality of the walls of the measuring vessel (heterogeneous nucleation), and the effect of light and of magnetic and electric fields.

In principle, the metastable zone width can be measured in two different ways—by isothermal and by polythermal measurements.

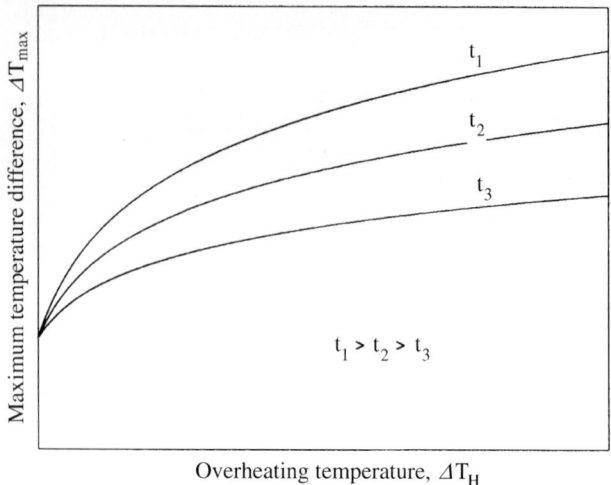

Figure 4.2.4 Schematic illustration of the effect of overheating temperature, ΔT_H, and of overheating time, t_i, on metastable zone width

4.2.1 Isothermal measurements

A saturated solution, agitated in a vessel, is supersaturated (for example, by cooling) at a high rate until a required supersaturation is reached at time t_0. This supersaturation, corresponding to the undercooling ΔT_2, is then maintained until the first crystals appear at time t_2. The time t_0 is usually

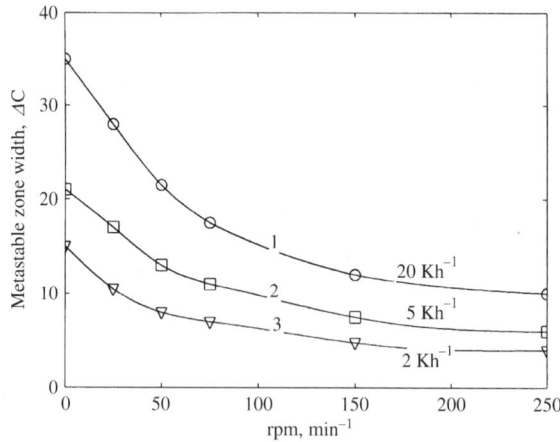

Figure 4.2.5 Effect of agitation on the metastable zone width in aqueous solution of $NaNO_3$[64]

160

not negligible compared to t_2. If, in the first period, the supersaturation was produced at rate $\dot{T}$, it follows that:

$$t_o = \frac{\Delta T_2}{(-\dot{T})} \tag{4.2.2}$$

Combining the power law:

$$B = k_N \Delta w^m \tag{4.2.3}$$

and the equation relating the induction period t_N to the nucleation rate:

$$B = \frac{K}{t_N} \tag{4.2.4}$$

we find[126]:

$$t_N = \frac{t_o}{(m+1)} + (t_2 - t_o) \tag{4.2.5}$$

The induction period at constant supersaturation $(t_2 - t_o)$ has to be increased by a portion corresponding to a fraction of the cooling period. In many cases, $m \approx 3$ so $t_o/3$ may be taken as a zeroth approximation. As an example, the induction periods of aqueous solutions of KCl (Figure 4.2.6) show that the scatter of experimental data is rather large. This is often found and arises from the stochastic character of nucleation. This data are plotted on logarithmic scales corresponding to the equation[126]:

$$\frac{1}{m} \log t_N = \frac{1}{m} \log \frac{K}{k_N} - \log \frac{dw^*}{dT} - \log \Delta T \tag{4.2.6}$$

(compare this with Equation (4.2.8)).

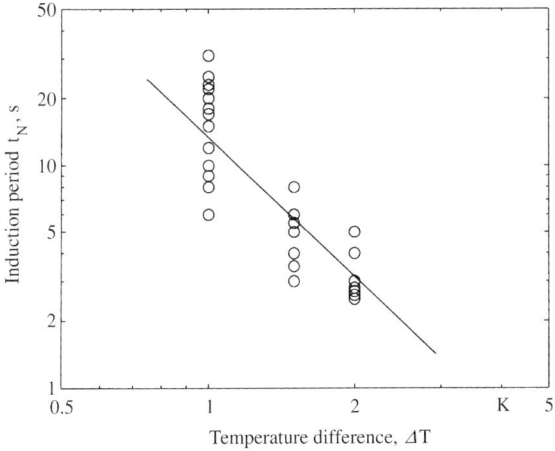

Figure 4.2.6 Induction periods of aqueous solution of KCl[126]

In salting-out crystallization or in precipitation with instantaneous addition of the second solution, the mixing time of the solutions t_o is very short and may be neglected. The supersaturation of the solution with sparingly soluble substances is[127]:

$$\Delta w = \frac{m_1}{m_0} - w_1^* \qquad (4.2.7)$$

and it follows that:

$$\log t_N = \log\left(\frac{K}{k_N}\right) - m \log \Delta w \qquad (4.2.8)$$

4.2.2 Polythermal methods

The polythermal method consists of measuring the metastable zone width at different cooling rates[64,128]. Assume that, at least in the initial stages of nucleation (on attaining the border of the metastable region), the mass nucleation rate equals the rate of production of solution supersaturation,

$$\left(\frac{dw^*}{dT}\right)\left(\frac{-dT}{dt}\right) = k_N \left(\frac{dw^*}{dT \Delta T_{max}}\right)^m \qquad (4.2.9)$$

Taking logarithms and rearranging yields a relationship between the independent variable, $\log(-\dot{T})$, and the dependent variable, $\log \Delta T_{max}$:

$$\log \Delta T_{max} = \frac{1-m}{m}\log\left(\frac{dw_{eq}^*}{dT}\right) - \frac{\log k_N}{m} + \frac{1}{m}\log(-\dot{T}) \qquad (4.2.10)$$

according to which the dependence of ΔT_{max} on $(-\dot{T})$ is linear on a logarithmic plot with a slope equal to $1/m$. An example of this plot is shown in Figure 4.2.7 on page 163.

The width of the metastable zone is usually measured by cooling a just-saturated solution at a constant rate until the first visible crystals are formed. It is not possible to exactly define the time of crystallization, as the size of the first visible particle depends on the observation method (for example, on the experimenter, apparatus, system, observed quantity, and so on). However, as the dependence of the nucleation rate on the supersaturation is very marked near the limit of the metastable region, the scatter in data measured under similar conditions should not be greater than that corresponding to the stochastic character of the process.

The size of just-visible crystals is usually between 10 and 100 μm; so critical nuclei must first grow to an observable size. If we take this growth into consideration[64,130], we obtain an equation almost identical to Equation (4.2.10) but with one change; the constant k_N has to be replaced by another constant

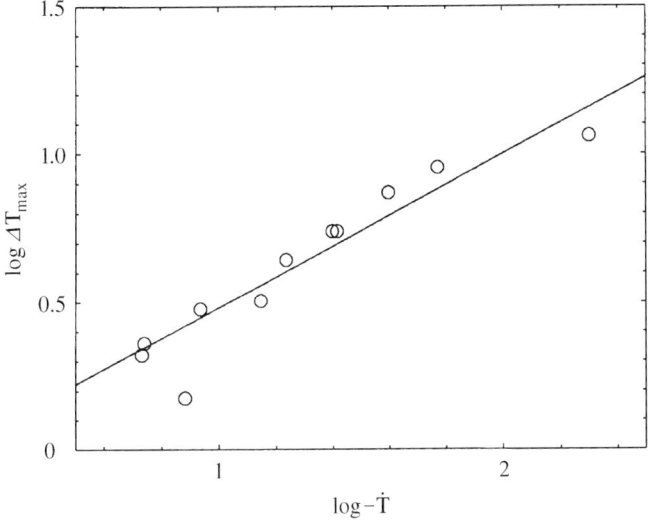

Figure 4.2.7 Linearized plot of the metastable zone width of $CuSO_4 \cdot 5 \ H_2O$[129]

which is a complex function of nucleation and growth parameters. For example, the apparent nucleation order, m, becomes a function of the true nucleation order, n, and of the growth order, g:

$$m = \frac{(3g + n + 4)}{4} \qquad (4.2.11)$$

The linear character of the dependence with the slope $1/m$, however, remains unchanged.

A suitable apparatus is shown schematically in Figure 4.2.8 (see page 164). The solution is kept in an Erlenmayer flask, equipped with a magnetic stirrer (m) and a thermometer. This thermometer may allow the mechanical operation of a constant cooling rate[131] or, better, will be a platinum resistance thermometer measuring the actual temperature of the solution and comparing this with a programmed temperature by a temperature programmer[128,132] or computer (R). This programmer operates a heating device, for example, an infrared lamp (L), that maintains the desired temperature of the solution. After cooling the solution to a temperature at which a sufficient number of tiny crystals are formed, the temperature is very slowly raised until the last crystal dissolves (the final rate of temperature rise being about $0.1 \ K \ h^{-1}$). This final temperature is considered to be the temperature of solution saturation, T*. Before the actual measurement of the width of the metastable region, the solution is maintained at a temperature about 0.5 K above the saturation temperature for about

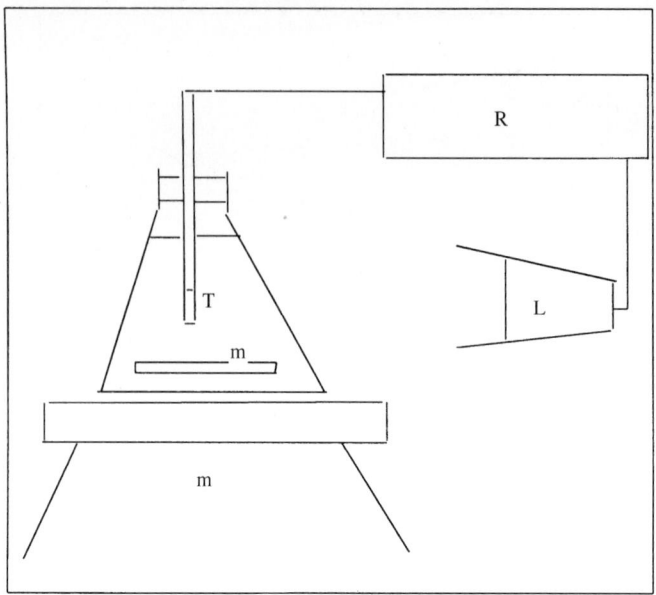

Figure 4.2.8 Apparatus for metastable zone width measurement

30 minutes. Then the solution is cooled at a constant rate of say 2, 5 or 20 K h^{-1} and the temperature at which the first crystals appear is recorded. The difference between this temperature and T* corresponds to the maximal undercooling, ΔT_{max}. By repeating the measurement several times for different cooling rates we obtain pairs of values of $(-\dot{T})$ and ΔT_{max} which can be correlated using Equation (4.2.10).

Appearance of the first crystals can be observed either by the naked eye in incident light (or using the Tyndall effect), by the change of intensity of transmitted light (which is less sensitive), or by monitoring some physical properties of the solution such as the refractive index, density or conductivity. An example of a conductivity measurement is shown schematically in Figure 4.2.9 (see page 165). We start with the suspension of tiny crystals at point (A). Heating of the suspension (dashed line) is accompanied by crystal dissolution until the last crystal is dissolved (B) and the solution is then slightly overheated (C). At this point cooling is commenced (solid line). At temperature T_N (point D) nucleation starts, leading to a rapid decrease in solution concentration. In the region of point (E), sufficient crystal surface has been formed to ensure the removal of supersaturation by growth (line E–A) and the measurement can be repeated.

Taking into consideration various modes of nucleation, we can carry out the measurements[133]:

164

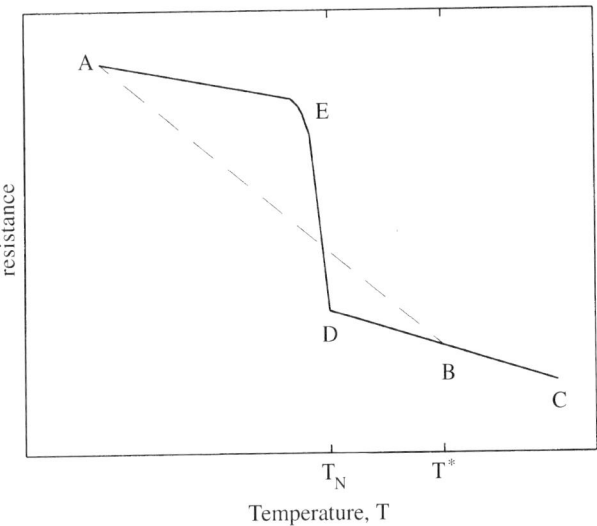

Figure 4.2.9 Conductivity–temperature plot of the metastable zone width measurement

- in the absence of crystals (primary nucleation, presumably heterogeneous);
- with a single seed immovably fixed in the stirred solution (secondary nucleation–catalytic mode); and
- with a few larger crystals moving over the vessel base (contact or collision nucleation).

Example
The metastable zone width of aqueous solutions of ammonium aluminium sulphate saturated at ∼ 40°C was measured by the method described above[133]. Three series of measurements were performed: (a) in the absence of solid phase; (b) with a single crystal, 3–5 mm in size mounted on a stationary holder immersed in the solution; and (c) with the same crystal moving freely on the bottom of the vessel with a magnetic stirrer rotating at a speed of about 200 rpm. The results are given in Table 4.2.1.

Treatment of the data with the method of least squares using Equation (4.2.10) gives the following equations:

series (a): $\log \Delta T_{max} = 0.3106 + 0.3327 \log(-\dot{T})$
series (b): $\log \Delta T_{max} = 0.2286 + 0.3009 \log(-\dot{T})$
series (c): $\log \Delta T_{max} = 0.0904 + 0.3372 \log(-\dot{T})$

Table 4.2.1 Results of metastable zone width measurements for ammonium alum expressed as ΔT, K

$-\dot{T}$ (K h^{-1})	Series (a)	Series (b)	Series (c)
2	2.8	2.1	
	2.55	2.0	
	2.35	2.3	
	2.75		
	2.52		
	2.9		
	2.4		
	2.6		
	2.5		
	2.64		
5	3.6	3.2	2.2
	3.35	2.3	2.0
	3.7	2.7	2.05
	3.3	3.0	1.9
	3.5	2.5	
	3.2		
	3.4		
	3.45		
10		3.1	2.85
			2.95
			2.8
			2.9
20	5.7		
	5.6		
	5.8		
	5.4		
	5.5		
25		4.8	3.4
			3.5
			3.7
			3.55

so that the nucleation parameters calculated from these equations using Equation (4.2.10) are (for $dw^*/dT = 0.09 \, K^{-1}$):

series (a): $m = 3.01$, $k_N = 0.00405$
series (b): $m = 3.32$, $k_N = 0.0130$
series (c): $m = 2.97$, $k_N = 0.0170$

The results are depicted in Figure 4.2.10 (see page 167).

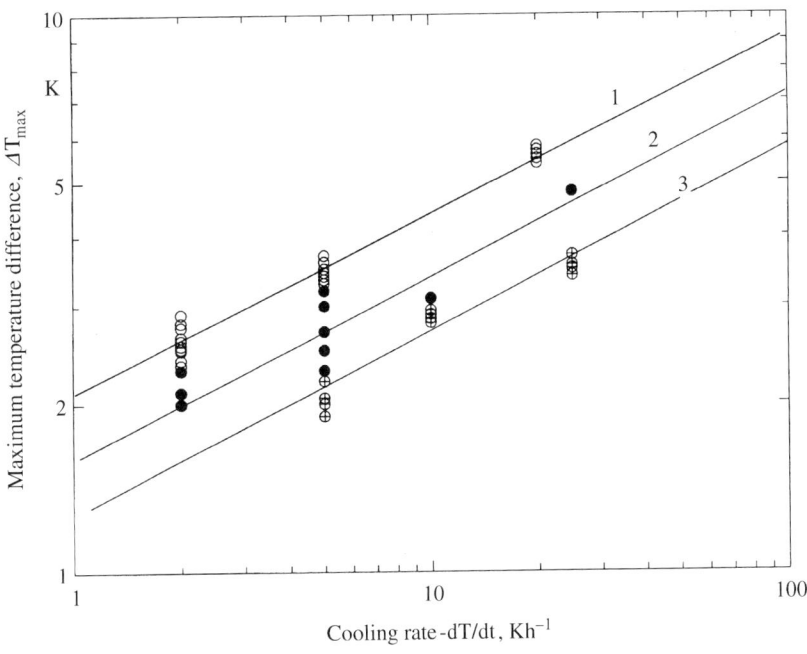

Figure 4.2.10 Metastable zone width of ammonium alum

Salting-out precipitation of inorganic salts from aqueous solutions by addition of an organic solvent miscible with water has attracted increasing attention. Metastable zone width measurements[134] can give valuable information for a process design.

To a system consisting of a salt (1) and a unit mass of water (0) which is just saturated, a second solvent is added at a constant rate $\dot{m}_2$. The addition of the second solvent proceeds by drops until the first crystals appear. If the solubility of the solute is affected by the second solvent according to:

$$w_1^* = w_{10} - K_{12}w_2 \qquad (4.2.12)$$

then the maximum supersaturation at the moment of precipitation of the first crystals, after having added m_2 of the second solvent, is:

$$\Delta w_{max} = \frac{K_{12}m_2(1 + w_{10})}{m_{10}} = K_{12}w_2 \qquad (4.2.13)$$

By analogy with the cooling case, we can show:

$$\log m_2 = \left[\frac{1 - m}{m}\log\frac{K_{12}}{m_{00}} - \frac{1}{m}\log k_N\right] + \frac{1}{m}\log \dot{m}_2 \qquad (4.2.14)$$

167

Table 4.2.2 Salting-out of magnesium sulphate by methanol

$\dot{m}_2 \, 10^6 \, (kg\,kg^{-1}\,s^{-1})$	$m_2 \, 10^3 \, (kg\,kg^{-1})$	m_2 calc. $10^3 \, (kg\,kg^{-1})$
21.34	249.8	235
42.69	274.0	294
85.38	363.8	369
106.7	390.5	397
213.4	518.7	497

So the plot of the mass of second solvent m_2 at the time of appearance of crystals plotted against the addition rate of the second solvent, $\dot{m}_2$ gives a straight line with the slope $1/m$.

Example

Methanol was used as the second solvent for salting-out $MgSO_4 \cdot 7\,H_2O$ from its aqueous solution[134]. Solution containing $m_0 = 63.25\,g$ of water and $m_1 = 66.75\,g$ of the hydrate at 20°C was placed into the apparatus and pure methanol added at a constant rate $\dot{m}_2$ until the first crystals appeared. The elapsed time and the mass m_2 were recorded. Experimental and smoothed data are presented in Table 4.2.2 and shown in Figure 4.2.11.

Starting from the Becker-Döring equation, we can relate the metastable zone width measurements to the classical nucleation parameters[64,130]:

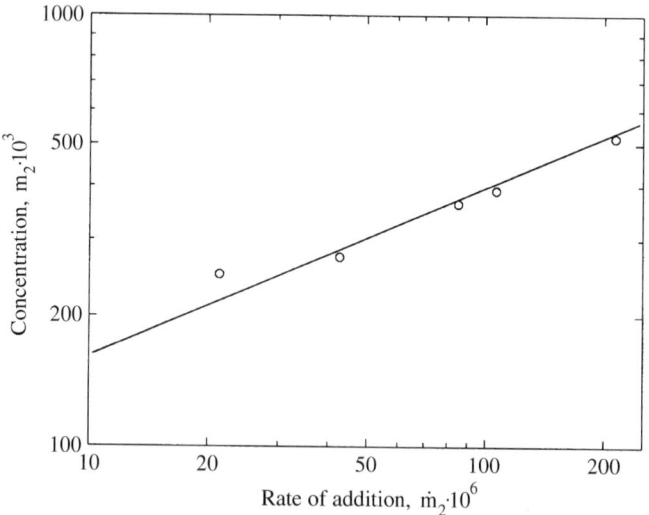

Figure 4.2.11 Metastable zone width of $MgSO_4$ salted-out by methanol[134]

the number of particles forming a critical nucleus is:

$$N^+ = \frac{n \, w^*}{\Delta w_{max}}$$

(4.2.15)

the size of the critical nucleus:

$$L^+ = 11.84 \left(\frac{N^+ M}{\alpha \rho_c} \right)^{1/3} \cdot 10^{-10} \quad (m)$$

(4.2.16)

and the specific surface energy of the critical nucleus:

$$\gamma_{CL} = 12471 \left(\frac{\alpha \rho_c T}{\beta M} \right) L^+ \frac{\Delta w_{max}}{w^*} \quad (J \, m^{-2})$$

(4.2.17)

Due to heterogeneous or even secondary nucleation, the specific surface energy obtained by this method is usually lower than the true value.

4.3 Rates of primary nucleation

Rates of primary nucleation are mostly determined by reaction or drowning-out crystallization. In general precipitation is a fast crystallization process carried out at high supersaturation. As a result of this high supersaturation, and the resulting high population density, nucleation is always accompanied by agglomeration and growth with the result that it is difficult to extract reliable nucleation rates from experimental results. It is therefore advisable to consider first some fundamentals of precipitation or reaction crystallization.

4.3.1 Fundamentals

In a precipitation process, two or more reactants (A, B) are mixed in a liquid environment and form one or more products (C, D). One of the products is sparingly soluble in the respective solvent and subsequently crystallizes:

$$A + B \rightarrow C \downarrow + D$$

(4.3.1)

As the particle size distribution of the drowned-out phase is an important specification of the product, the system needs to be described using not only the materials and enthalpy balances, but also a population balance. This relates the size L of the particles and the population density n. The general form of this equation is[10]:

$$\frac{\partial n}{\partial t} + \frac{\partial (Gn)}{\partial L} + D(L) - B(L) + n \frac{\partial (\ln V)}{\partial t} + \sum_k \frac{n_k \dot{V}_k}{V} = 0$$

(4.3.2)

The population density n is the number of particles in the size range ΔL per unit volume V of the slurry. Many of the parameters in Equation (4.3.2) have been described in earlier sections, for example the crystal growth rate G, the birth rate B, and the death rate D. In this section we focus on the governing parameter for precipitation processes. This is the number of particles formed, ∂N_{tot}, based on the time period of their formation ∂t and the volume of the slurry ∂V in which they were generated; this is the nucleation rate. If we only consider nucleation and do not implement effects like growth and agglomeration, the birth rate B is not size dependent and is equal to the nucleation rate:

$$B = \frac{\partial N_{tot}}{\partial t \, \partial V} \tag{4.3.3}$$

A predictive calculation of the birth rate B_{hom} for homogenous nucleation was derived by Kind and Mersmann[135] based on the classical theory of nucleation[136] and valid for any crystallizing system:

$$B_{hom} = 1.5 \, D_{AB}(c \cdot N_A)^{7/3} S^{7/3} \sqrt{\frac{\gamma_{CL}}{kT}} \cdot V_m$$

$$\times \exp\left[-\frac{16\pi}{3} V_m^2 \left(\frac{\gamma_{CL}}{kT}\right)^3 \frac{1}{(v \ln S)^2}\right] \tag{4.3.4}$$

Although in principle experiments are no longer necessary to obtain nucleation kinetics in the homogeneous region with this predictive calculation, several uncertainties in Equation (4.3.4) must be overcome. The coefficient D_{AB} is assumed to be independent of concentration which is not valid but not too decisive. The interfacial energy γ_{CL} between the crystal and the mother liquor is a physical property which cannot be directly measured. This material specific contribution can be calculated from Equation (4.1.13)[137]. However, the uncertainty of the factor K ($0.31 < K < 0.414$) leads to great differences of nucleation rates and metastable zone widths because of the expression $B_{hom} \sim \exp(-const(\gamma_{CL})^3)$. Also γ_{CL} strongly depends on solution conditions like suspension pH and impurities. Ionic dissociation also plays an important role, where the number of ions v and the degree of dissociation α need to be known. Most importantly, the supersaturation S depends on activities and therefore includes the non-idealities of the system; this has to be calculated exactly (see Equation (2.2.3)). Due to the low concentrations of the sparingly soluble substances, neglect of the activity coefficients (proposed in Equation (2.2.4)) can lead to large errors. Particularly for multi-component systems and electrolyte systems containing organic substances, these uncertainties restrict the predictive value of Equation (4.3.4). However, it well decribes the strong dependence of the nucleation rate B on the supersaturation S, shown in Figure 4.3.1 (see page 171), and may serve to estimate the order of

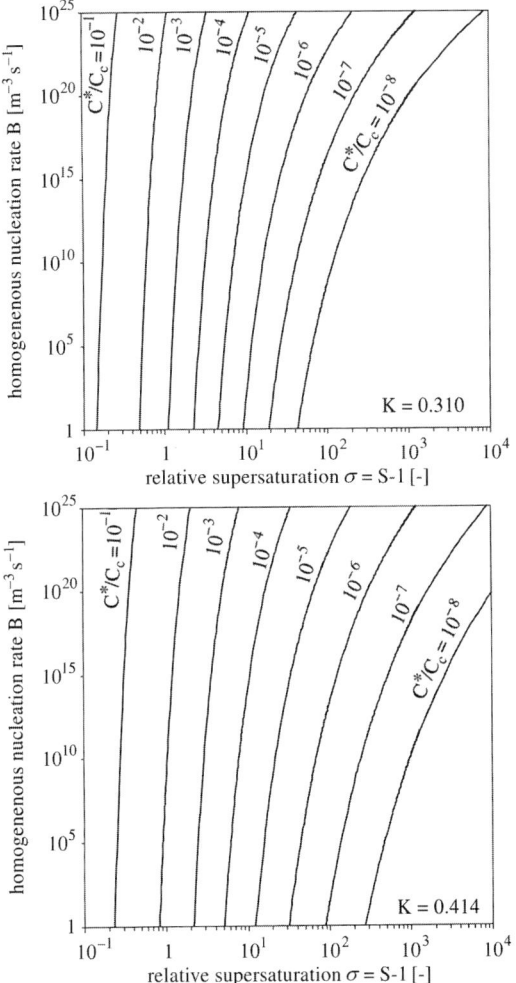

Figure 4.3.1 Calculated rates of homogeneous nucleation versus supersaturation for various dimensionless solubilities C^*/C_c, see Equation (4.3.4) (valid for $D_{AB} = 1.5 \cdot 10^{-9}\,m^2\,s^{-1}$; $v = 2$; $c_c = 21\,kmol\,m^{-3}$; $T = 293\,K$ and $\gamma_{CL} = K\,kT/d_m^2\,\ln(c_c/c^*)$)

magnitude. In such diagrams the metastable zone width σ_{met} is well defined for a given nucleation rate.

Besides those restrictions on a predictive calculation of homogenous nucleation, it is even more difficult to account for the presence of foreign particles which leads to heterogeneous nucleation. Despite recent efforts in modelling heterogeneous nucleation[125], the number of ill-defined parameters increases, which in turn makes predictive calculations even more difficult. Therefore, an experimental determination of the nucleation rate B is essential and in many cases still the most precise method. Different techniques and experimental set-ups will be discussed and explained in detail. First, however, there are some general requirements to be noted to ensure that the experimental study leads to nucleation rates that can be scaled up with reasonable confidence.

4.3.2 General requirements for an experimental technique

(a) Agglomeration

On examining Equation (4.3.2), and presuming that the growth rate is already determined, the remaining unknown parameters are the birth rate B and the death rate D. From experimental evaluation, there is no distinction possible between D and B, as the only information we gain is the right hand side of Equation (4.3.3), which in arbitrary conditions reflects the combination of both terms. So when determining the nucleation rate experimentally, it is essential to avoid any mechanism affecting the total number of the particles other than the nucleation process. In precipitation processes, attrition is usually negligible as the particles are small, but agglomeration might become the dominating effect if conditions are not carefully chosen. If the precipitation is run in the colloidal domain (particles less than about 1 μm in size) the agglomeration is perikinetic and proportional to the square of the population density n[124]:

$$\frac{\partial n}{\partial t} \sim -W \cdot n^2 \qquad (4.3.5)$$

where W is an effectiveness factor depending mainly on material specific Van der Waals attraction of two particles and the electrostatic repulsion[138]. In order to keep agglomeration negligible, the suspension density should be kept below about 0.1% in volume. For some materials, like barium sulphate, the effectiveness is quite low so agglomeration is less dominant.

Without going into detail, we propose two possibilities to ensure electrostatic repulsion in electrolytic suspensions, which is especially important for materials with a higher tendency to agglomerate. Firstly, the ionic strength, which is the total concentration of all ions in the solution, should be kept as low as possible, so any inert electrolyte should be avoided in the solution. Secondly, the suspension pH strongly controls the electrostatic repulsion. To use this phenomena the isoelectric point must be known, which is the material specific pH for which the repulsive electrostatic potential is zero. The further the suspension pH is from this isoelectric point, the more agglomeration can be avoided. If the role of agglomeration is not taken into account, the measured nucleation rate can be the result of nucleation and agglomeration.

(b) Mixing

Mixing phenomena play a major role in almost every precipitation reactor, whether on the industrial or laboratory scale. Firstly, the reactants have to come into contact on a molecular level to build up the supersaturation which leads to crystallization. This mass transfer process can be controlled by the local power input. When focusing on nucleation, the build up of supersaturation is critical, as the chemical reaction as well as the nucleation itself are usually very fast processes. It is only in the case where mixing is uniform that the gained kinetic data do not depend on mixing, but only on thermodynamic parameters. Mixing times on the molecular level τ_{micro} (called micromixing) correspond to the local specific power input ε into the system, the kinematic viscosity ν_L and the Schmidt number Sc. Equation (4.3.6) is valid for a degree of segregation $S_{seg} = 0.1$[88]:

$$\tau_{micro} \approx 5 \sqrt{\frac{\nu_L}{\varepsilon}} \ln Sc \qquad (4.3.6)$$

In stirred vessels the spatial distribution of the specific power input ε is very inhomogenous, with high values close to the stirrer but large dead zones with only poor mixing on the molecular level near the surface and walls. The inhomogeneity increases with increasing vessel volume. This leads to inhomogeneity of the supersaturation and nucleation then takes place at a locally undefined supersaturation in an undefined volume. The evaluation of nucleation rates becomes impossible bearing in mind Equation (4.3.3). So when using stirred vessels, it is recommended that vessels of less than $0.1 \, dm^3$ are used, with a power input high enough to obtain micromixing times of 10^{-4} s.

T- or Y-mixing devices overcome many of these difficulties, as they have a very confined zone of mixing with the possibility of high local power inputs. Although these devices are less used on an industrial scale as they tend to block, this does not matter on the laboratory scale with short operating times. To calculate micromixing times of these T- or Y-mixers, the specific power input ε can be expressed through the local pressure drop Δp, the volumetric flow rate $\dot{V}$ and the mass of the slurry m in the volume $\int \partial V$:

$$\varepsilon = \frac{\Delta p \cdot \dot{V}}{m} = \frac{\Delta p \cdot \dot{V}}{\rho_{sus} \int \delta V} \qquad (4.3.7)$$

Here ρ_{sus} is the density of the suspension.

Figure 4.3.2 on page 174 illustrates the design of a Y-mixer. The local pressure drop in the mixing chamber can be estimated from the ratio of the fluid

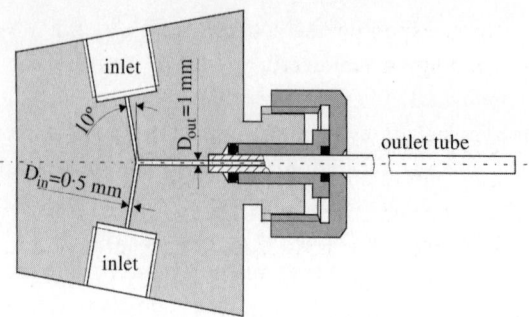

Figure 4.3.2 Design of a Y-mixer

velocities in the tubes leading to the mixing chamber, w_{in}, and the outlet tube w_{out}[139]. The specific power input is given by:

$$\varepsilon = \xi \, \frac{w_{in}^3}{2D_{in}} \qquad\qquad (4.3.8)$$

with $\xi = 2$ for a velocity ratio of $w_{out}/w_{in} = 1$ and $\xi = 5$ for a velocity ratio of $w_{out}/w_{in} = 2$. With a local pressure drop of $2 \cdot 10^5$ Pa, such T- or Y-mixer devices can achieve micromixing times of $2 \cdot 10^{-4}$ s, which is fast enough for the majority of precipitation reactions. Care has also to be taken that the flow in the outlet tube is turbulent $((w_{in} \cdot D_{in}/v) > 2300)$. The device shown in Figure 4.3.2 allows a variation of the residence time in the outlet tube by using tubes of different length.

4.3.3 Integral determination with T- or Y-mixers

In Equation (4.3.3) the nucleation rate is defined as the differential change of the number density with time. The determination of this differential change requires a particle analysis system with the ability to count particles online, which is (at least for nuclei of a few nanometers) not yet available. However, it is possible to specify a precise time interval within which nucleation takes place, and to determine the total number of nuclei generated in this time. This is the integral method of determination.

T-mixer and Y-mixer devices give the ability to specify precisely the time during which nucleation takes place through the residence time in the outlet tube, provided that the suspension is quenched directly at the end of the outlet tube of the T-mixer. As will be discussed later, it is important to ensure that the supersaturation is nearly constant in the outlet tube. The advantage in using these devices is the very intensive mixing in a very confined volume.

Two slightly different experimental techniques are discussed below, with the most suitable depending on the specifications of the substances investigated.

(a) Experimental technique using a particle counter

At present, particle analyzers capable of counting crystals in a slurry have a lower size limit of detection of about 1 μm. This is not small enough to detect the nucleation event itself. For this reason an experimental set-up is recommended which consists of a nucleator followed by a growth vessel, where the nuclei grow at moderate supersaturations to a detectable size. A typical experimental set-up is shown in Figure 4.3.3.

The experimental concept is to separate the process of nucleation in the T-mixer from growth of the generated particles in the growth vessel. The experiments are run discontinuously with a certain reactant volume fed through the T-mixer into the vessel at the beginning of the experiment. In the T-mixer rapid mixing occurs due to high specific power input in the mixing zone. The supersaturation reaches its nominal value at the entrance of the outlet tube if the times for the micromixing and for the chemical reaction are very short in

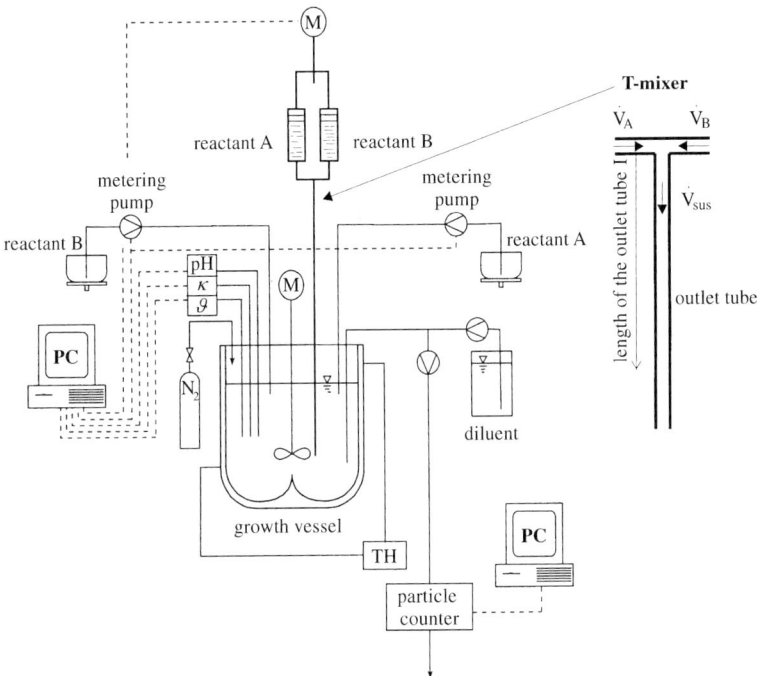

Figure 4.3.3 Experimental set-up[125,140,141]

comparison to the residence time of the slurry in the outlet tube. Nucleation thus takes place in a precisely defined volume at a precisely defined level of supersaturation. At the end of the outlet tube the size of the generated particles is in a range of nanometers.

The suspension is instantly diluted upon entering the agitated vessel, the nucleation process stops but the crystals continue to grow. The supersaturation in the vessel is kept at such a low level (typically $6 \leq S_a \leq 10$) that moderate growth occurs, but at the same time the nucleation rate is negligibly small. Figure 4.3.4 illustrates the course of such an experiment.

The nucleation rate B is calculated from the total number of particles N_{tot} generated in the tube according to:

$$B = \frac{N_{tot}}{\tau \cdot V_{tot}} = \frac{4\dot{V}}{\pi D_{out}^2 l_{out}} \cdot \frac{N_{tot}}{V_{tot}}$$

(4.3.9)

with V_{tot} representing the volume of the reactant solution fed through the T-mixer, and l_{out} the length of the outer tube.

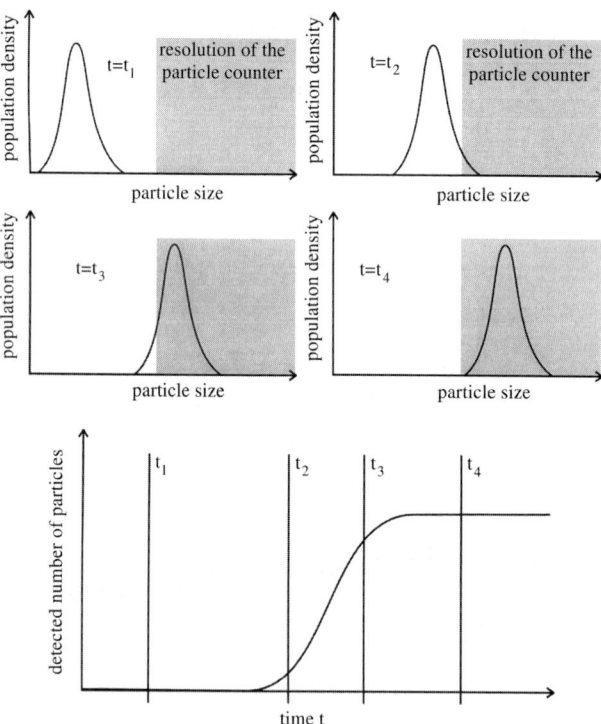

Figure 4.3.4 Course of a nucleation experiment[125]

The residence time τ of the slurry in the T-mixer is only a rough measure of the actual nucleation time t, since the build up of supersaturation can take a certain period of time. It is also necessary to check, for every experiment, whether the nominal value of the supersaturation remains practically constant throughout the T-mixer. If the supersaturation decreases significantly due to nucleation and growth processes the evaluation of nucleation rates becomes uncertain. As an example, Figure 4.3.5 shows the calculated course of supersaturation along the outlet tube of the T-mixer (length $= 0.2$ m) versus the distance from the mixing chamber for nucleation experiments with barium sulphate[125].

The calculations show that for barium sulphate, the supersaturation remains almost constant along the crystallizer tube up to a value of $S_{a,o} \approx 500$, while for $S_{a,o} \geq 800$ the supersaturation decreases substantially. For very high initial supersaturations (i.e. $S_{a,o} = 1500$) the decrease is so rapid that nucleation rate measurements would be unreliable.

In the following, the evaluation of an experiment is shown for the set-up of Figure 4.3.3 using barium sulphate as a model system. $BaSO_4$ is precipitated from the reactants $Ba(OH)_2$ and H_2SO_4:

$$Ba(OH)_2 + H_2SO_4 \rightarrow BaSO_4 \downarrow + 2H_2O \qquad (4.3.10)$$

An advantage of this system is that water is the by-product of the chemical reaction, which makes it easy to control the supersaturation by measuring the

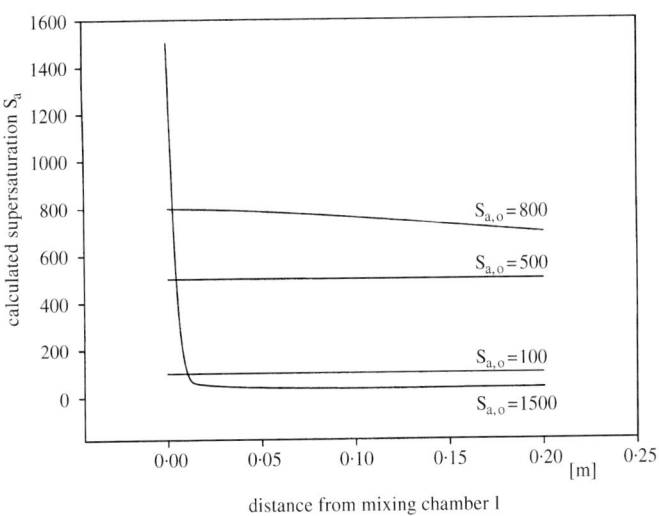

Figure 4.3.5 Course of supersaturation S_a for four different initial values $S_{a,o}$ calculated for barium sulphate[125]

electric conductivity and the pH. To avoid the formation of $BaCO_3$, the apparatus is kept under a nitrogen atmosphere.

Experiments can be carried out over a wide supersaturation range ($S_a \leq 925$, the maximum value being given by the solubility of the reactants) and the supersaturation S_a is calculated based on activities. The reactants are added stoichiometrically (this is not a prerequisite for the experiments). Ultra-pure water which is first distilled and then passed through a Milli-Q Millipore water purifying system serves as the solvent. The precipitation temperature is controlled at a fixed value (for example, $T = 298\,K$). A Pamas PMT-2100 particle size analyser is used to detect and count the generated particles.

The details of the experimental set-up are summarized in Table 4.3.1 and important experimental details are as follows:

Before the experiment:

- The supply tanks are filled with barium hydroxide solution and diluted sulphuric acid, respectively.
- The growth vessel ($V = 0.8\,dm^3$) is filled with saturated $BaSO_4$ solution. The amount is chosen in such a way that the growth vessel is full sufficiently after the reactants have been added through the T-mixer.

Start of the experiment:

- The electric conductivity of the saturated solution in the growth vessel is measured as a reference value for supersaturation control.
- The reactants are fed through the T-mixer into the growth vessel ($w_{in} = 8.6\,m\,s^{-1}$) and the slurry entering the stirred vessel is rapidly diluted as a result of the high stirrer speed.

During the experiment:

- The supersaturation S_a in the growth vessel is monitored by measuring the electrical conductivity. If the supersaturation falls below a certain value, highly diluted reactant solutions are fed stoichiometrically into the vessel.

Table 4.3.1 Main data of the experimental set-up

Length of the T-mixer	$0.2\,m$
Inner tube diameter	$0.8\,mm$
Reactant volume, V	$\approx 800\,ml$
Feed rate of the reactants in the T-mixer	$8.6\,m\,s^{-1}$
Slurry volume in the growth vessel	$800\,ml$
Supersaturation S_a in the growth vessel (based on activities)	$6 \leq S_a \leq 10$

• An extremely small amount of the slurry is constantly removed from the vessel and fed through the particle analysing system to monitor the particle concentration.

After the experiment:

• The maximum number of particles is taken as corresponding to the number of nuclei N_{tot} created in the T-mixer.
• The nucleation rate B is calculated according to Equation (4.3.9).

The applicability and reliability of this technique to grow particles into the resolution range of a particle counter depends strongly on the growth rate of the investigated substance and its tendency to agglomerate. The time necessary to grow barium sulphate nuclei precipitated in a T-mixer to the size range above 1 μm is about 300 seconds[141]. Thus this period can easily reach some hours if the growth rate is low and the particles exiting the T-mixer are very small due to higher nucleation rates[122]. Here agglomeration can become dominant with negative effects on the reliability of the kinetic data as mentioned before. For barium sulphate these effects can lead to significant deviations[125].

(b) Experimental technique using particle size analysis

Instead of growing the particles over a range of up to three decades to enter the resolution of currently available particle counters, the nucleation rate can be determined by measuring the particle size distribution and the total mass precipitated. This technique is particularly favoured when the initial particles are small, the growth rates are low, and agglomeration cannot be suppressed sufficiently. Most available devices for particle size analysis do not reach the size range of the nuclei, which are only a few nanometers, although recent developments are achieving significantly smaller sizes. For example, ultra-centrifuges can analyse down to 10 nm, but analysing times become very long. More promising are quasielastic (or dynamic) light scattering (QUELS) devices, such as photon correlation spectroscopy (PCS) and homodyne detection devices. Both allow detection over the size range from about 0.003 to 1 μm. Care has to be taken with PCS to keep the suspension density low and at a constant value. Both devices should be able to display reliably multimodal distributions, which are rather common in precipitation processes.

The experimental set-up for this technique is very similar to that described in Figure 4.3.3. A Y-mixer is used as a nucleator and the time for nucleation is specified precisely by quenching the suspension at the end of the tube.

179

The difference with this technique is that instead of supplying the vessel with a small but constant supersaturation to enable crystal growth, the precipitates are quenched in double distilled and purified water containing surfactants to hinder agglomeration and stabilize the colloid. This is important as the measurement times for the QUELS apparatus are around ten minutes and several samples should be taken in that time in order to enhance the statistical significance. A complexing agent can also be added to the diluting medium in order to block the remaining supersaturation.

Two typical experimental set-ups are shown in Figure 4.3.6. In Figure 4.3.6a the slurry is quenched in a stirred vessel at the zone of maximum energy dissipation rate, as is described in Figure 4.3.3. In Figure 4.3.6b the slurry is quenched with the aid of a second Y-mixer. This increases the intensity of quenching, but is also more difficult to operate. In both facilities the reactants are fed with a stepping motor driven double piston device, which enables exact control of the flow rates and the amount of reactants. The vessel in Figure 4.3.6a is $1\,dm^3$ in volume, is provided with baffles and a draft tube, and is mixed with a propeller stirrer at a mean specific power input of $\bar{\varepsilon} = 0.3\,W\,kg^{-1}$.

The course of an experiment is as follows.

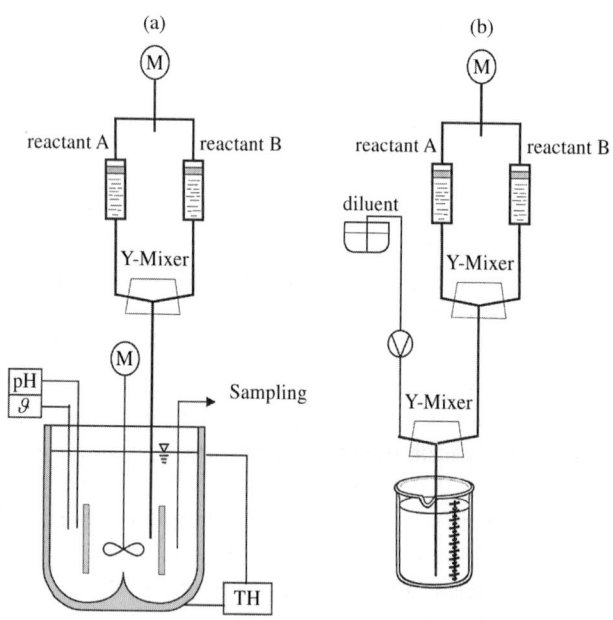

Figure 4.3.6 Experimental set-up[96,122]

Before the experiment:

- The reactants are prepared with double-distilled and ultrafiltered water in concentrations to achieve the required stoichiometry and nominal super-saturation in the Y-mixer.
- The diluent is prepared with double-distilled and ultrafiltered water and the stabilizing surfactant. The concentration of the surfactant should be well below the critical micelle concentration, since the micelles may be in the same size range as the nuclei and hence may also be detected in the particle size analysis. The dilution should be performed at iso-pH. It is convenient to run an experiment beforehand without dilution, measure the pH and adjust the pH of the diluent with the aid of common buffers to this value. The use of a complexing agent is also recommended to ensure that no further growth or nucleation affects the result. Fluoride ions for example are fast complexants for a variety of cations. The concentration of the complexant has to be chosen carefully; high enough to block the remaining supersaturation, but not too high that precipitates might dissolve.
- The vessel is filled with the diluent. The volume of diluent $V_{Diluent}$ is chosen so that the dilution ratio $V_T/V_{Diluent}$ is between 1:10 and 1:100, depending on the initial supersaturation in the Y-mixer and also bearing in mind that the suspension density after dilution must be in the appropriate range for the particle analyser. The vessel may be cooled down just beyond the freezing point to reduce further the agglomeration and crystallization kinetics when the product is quenched.
- The pistons are filled with the reactants and heated to the desired temperature.

Start of the experiment:

- When employing the set-up using the vessel, the stirrer speed is adjusted to say 600 rpm, which at the given geometry (see Table 4.3.2 on page 182) corresponds to a mean specific power input of $0.3 \, W \, kg^{-1}$. When using the second Y-mixer for quenching, the pump for the diluent is started simulta-neously with the stepping motor. The flow rate is set to correspond to the flow rate of the suspension leaving the first Y-mixer to obtain the specified dilution rate.
- The stepping motor is set to the required flow rate and feed volume V_T of the reactants and started. For the geometry with the Y-mixer given in Table 4.3.2, the reactant flow rates are set to $3.5 \, ml \, s^{-1}$, corresponding to velocities of $18 \, m \, s^{-1}$ in the inlet tubes and leading to calculated micromix-ing times below $10^{-4} \, s$ with a local specific power input $\varepsilon \approx 10^7 \, W \, kg^{-1}$.

Table 4.3.2 Main data of the experimental set-up

Max. volume of a piston	100 ml
Feed volume of reactants	$20\,\text{ml} \leq V_T \leq 200\,\text{ml}$
Flow rate of reactants	$3.5\,\text{ml s}^{-1}$
Diameter of Y-Mixer inlet tubes	0.5 mm
Diameter of Y-Mixer mixing chamber	1 mm
Diameter of Y-Mixer outlet tube	2 mm
Length of Y-Mixer outlet tube	variable
Residence time in the Y-Mixer outlet tube	$10^{-2}\,\text{s} \leq \tau \leq \infty$
Vessel dimensions (Diameter D, Height H)	$D = H = 0.1\,\text{m}$
Propeller stirrer diameter	0.058 m
Stirrer speed	600 rpm
Dilution ratio	$1{:}10 \leq V_T/V_{\text{Dilutant}} \leq 1{:}100$

When using the second Y-mixer for quenching, the pump for the diluting medium is started simultaneously.

End of the experiment:

- Five samples are taken for off-line analysis of the particle size distribution. Here a MICROTRAC© UPA150 apparatus, which is a homodyne QUELS device, has been used.
- Another sample is taken, ultracentrifuged at 15,000 g for an hour, and dried to determine the mass fraction m_T of precipitates in the suspension. This procedure is recommended, as a mass balance over the entire process using the initial concentrations of the reactants and the solubility does not reflect the unknown amount in the still prevailing or complexated supersaturation.
- When the density ρ_C of the product is unknown, due for example to a partly amorphous or polycrystalline precipitate, it also has to be determined. Possible apparatus are pycnometers like the MICROMERITICS© ACCU PYC 1330.
- The total number of particles nucleated within the experiment N_{tot} is calculated from:

$$N_{\text{tot}} = \frac{m_T(V_T + V_{\text{Diluent}})}{\alpha \cdot \rho_C \cdot L_{50}^3} \tag{4.3.11}$$

where α is the volume shape factor and L_{50} the determined volume median crystal size.

- The nucleation rate B is calculated according to Equation (4.3.9). The residence time τ of the slurry in the outlet tube of the Y-mixer is taken to correspond to the nucleation time t, and the same precautions have to be taken into account as are mentioned in Section 4.3.3a. The time for the build-up of the supersaturation is negligible, which is only a valid assumption when the conditions for the micromixing are kept in the range described above.

- The resulting nucleation rate is related to the nominal supersaturation $S_{a,0}$ calculated using activities from the initial concentrations, which is assumed to be constant throughout the entire outlet tube. The reliability of this assumption must be checked with the aid of a desupersaturation calculation, starting from the measured kinetic data. When the desupersaturation is found to be too high, the study should be performed at lower initial supersaturations (see Section 4.3.3a), or the residence time should be reduced using a shorter outlet tube.

The data of the experimental set-up are summarized in Table 4.3.2.

Whether the set-up with a second Y-mixer or the quench into a stirred vessel is the most appropriate depends on the dilution rate. At low dilution rates the set-up with the second Y-mixer has the advantage of good and intensive quenching, but this advantage becomes less important for increasing dilution rates. Also, the high pressure drop at high dilution rates is a drawback of the technique using a second Y-mixer, so the technique using a stirred vessel is more appropriate at high dilution rates.

The reliability of the nucleation rates measured using this technique is, ignoring agglomeration, restricted mostly through the accuracy in determining the total number of precipitated particles (Equation 4.3.11). Different measured quantities contribute to the error in this determination, from which the estimation of the median crystal size is the most important. As previously mentioned, a direct measurement of the particle number through a particle counter is preferable when the time to grow the precipitates is short and agglomeration is sufficiently small. In addition to the possibility of agglomeration at the diluted state, there is also the possibility that agglomeration in the T- or Y-mixer itself will restrict the accuracy of determining nucleation rates. Bi- and trimodal particle size distributions are a stong indication that agglomeration has taken place. To try to avoid agglomeration, the suspension density in the Y-mixer should be decreased, and so the experiment has to be run at lower supersaturation levels[122].

4.3.4 Desupersaturation experiments at constant volume

Looking at the population balance, Equation (4.3.2), the exploitation of nucleation experiments is made easier when the volume V is kept constant and when all inlet and outlet streams are set to zero. In batch desupersaturation experiments, the supersaturation diminishes from an initial value throughout the course of the experiment. This is the main advantage of this technique, as the kinetic data obtained from one single experiment relates to a large range of supersaturation and, overall, the determination of crystallization kinetics is less laborious. Different exploitation routes have been developed based on a moments analysis, a Laplace transformation, or a Fourier transformation of the particle size distribution[142,143]. Most of these techniques were developed for seeded crystallizers.

Compared to the MSMPR technique, the inhomogeneity through incomplete mixing that arises once the build-up of the initial supersaturation has taken place is less important, since the desupersaturation occurs in the entire volume of the crystallizer.

However, when trying to determine kinetics of primary nucleation (the governing mechanism in precipitators) in such batch experiments, the supersaturation has to be maintained at least above the critical value without adding seeds in order to achieve thermodynamically activated nucleation. But it is in this very unstable period of the experiment that the question of a homogeneous distribution of the supersaturation in the crystallizer is important. To achieve reliable kinetic data therefore, the experiment has to be performed for a batch time where the build-up of supersaturation is totally completed and only desupersaturation due to nucleation and crystal growth is prevailing (see Figure 4.3.7). Furthermore, to ensure that the study leads to nucleation rates from a primary mechanism, the supersaturation has to be above the critical value. It is difficult to meet these very specific demands, as nucleation is a rather fast process and the experiment has to be stopped before crystal-induced (secondary) nucleation mechanisms start to play a major role. To achieve kinetic data for lower values of supersaturation, however, the determination from desupersaturation experiments is a valuable method. The following procedure, based on a moments analysis of the population balance, leads theoretically to a complete description of the nucleation kinetics from one single experiment[144].

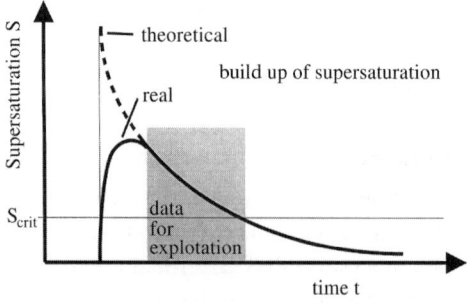

Figure 4.3.7 Course of the desupersaturation with batch time

The number of particles per volume of the slurry at a batch time t is determined from the zeroth moment m_0 of the population density distribution n:

$$m_0(t) = \int_0^\infty n(L, t)dL \qquad (4.3.12)$$

The nucleation rate B(t) is then determined from the difference between the zeroth moments in the small time interval Δt:

$$B(t) = \frac{\Delta m_0(t)}{\Delta t} \qquad (4.3.13)$$

When there are sufficient measured data of the supersaturation and the number density n in the exploitable time period available (see Figure 4.3.7 on page 186), the kinetics law can be derived from the data of only one single experiment. In practice the experiment should be repeated several times.

Although the experimental set-up for this technique is very simple, there are certain demands if the supersaturation measurement and the particle size analysis system are to work online and provide many measurements in order to achieve significant kinetic data. If the particle analysing system does not provide a population density distribution, but only a normalized volume distribution (such as laser diffraction and QUELS apparatus), a transformation from volume to number density has to be performed, whereby any error of the analysing system is tripled. The population density is obtained through calculation of the actual suspension density from a mass balance, taking into account the shape factor of the particles.

The resulting kinetic data depend strongly on the resolution of the particle analysis system. The time that elapses before the precipitates grow into the detectable size range means that the nucleation rate is always referred to a supersaturation prevailing some time after the generation of the nuclei. When trying to take this effect into account the calculation becomes very complicated and still more uncertain if the median growth rate for the entire population is used. This is especially so for very small particles when size dependent effects have to be considered[145]. Agglomeration can again be a severe problem and might become more dominant as the residence time is much higher.

4.3.5 MSMPR experiments

Although continuously-operated crystallizers with MSMPR are well-established for highly soluble substances and, due to the simple treatment of the population balance, are a valuable technique, they are not recommended for precipitation for several reasons. As mentioned above, the mixing in stirred vessels is highly inhomogeneous in terms of micromixing. As a result, one of

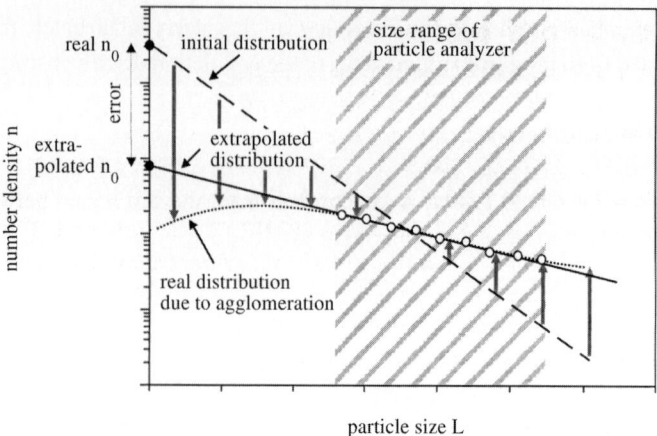

Figure 4.3.8 Evaluation of nucleation rates of precipitates according to the MSMPR model

the key assumptions in deriving the MSMPR equations, constant supersaturation in the entire volume and perfect mixedness of the slurry, is not met. When calculating the nucleation rates based on the entire volume of the suspension in the vessel, the calculation can be far from reality because the nucleation is likely to take place in a small but poorly defined zone close to the inlet of the reactants, where the supersaturation is also unknown.

Another source of error arises from agglomeration as the slurry consists of submicron particles at a rather high suspension density and the residence time is comparatively long. The presence of significant agglomeration decreases the total number of particles, and the determined nucleation rates refer to 'apparent nuclei', which are in reality agglomerates and not primary particles. If the particle size analysis system is restricted in resolution, this effect cannot be detected as is illustrated in Figure 4.3.8. Therefore, this technique involves much uncertainty and resulting experimental data are open to misinterpretation.

References and bibliography

1. Nyvlt, J., 1972, *Kristall und Technik*, 7, 12, K127
2. Mullin, J.W., 1973, *The Chem. Engr*, June, 316
3. Mullin, J.W., Söhnel, O., 1977, *Chem. Engng. Sci.* 32, 683
4. Söhnel, O., Garside, J., Jancic, S.J., 1977, *J. Crystal Growth* 39, 307
5. Söhnel, O., Garside, J., 1979, *J. Crystal Growth* 46, 238
6. Cardrew, P.T., Davey, R.J., Garside, J., 1979, *J. Crystal Growth* 54, 358
7. Mullin, J.W., 2001, *Crystallization, 4th ed.*, Butterworth-Heinemann (London, UK)
8. Burton, W.K., Cabrera, N., Frank, F.C., 1951, *Phil. Trans. Roy. Soc.* 243, 299
9. Van der Eerden, J.P., Bennema, P., Cherepanova, T.A., 1978, *Prog. Crystal Growth Characterization*, 1, 219
10. Randolph, A.D., Larson, M.A., 1988, *Theory of Particulate Processes, Analysis and Techniques of Continuous Crystallization*, Academic Press, Inc. (San Diego, CA, USA)
11. Bird, R.B., Steward, W.E., Lightfoot, E.N., 1960, *Transport Phenomena* Wiley (New York, USA)
12. Mersmann, A., 1988, Design of Crystallizers, *Chem. Eng. Process.* 23, 213–228
13. Mersmann, A., Angerhöfer, M., Gutwald, T., Sangl, R., Wang, S., 1990, General Prediction of Mean Crystal Sizes, *Proceedings of the 11th Symposium on Industrial Crystallization 1990, (ed. A. Mersmann)*
14. Garside, J., 1971, *Chem. Engng. Sci.* 26, 1425
15. Garside, J., Tavare, N.S., 1981, *Chem. Engng. Sci.* 36, 863
16. Garside, J., Phillips, V.R., Shah, M.B., 1976, *Ind. Eng. Chem. Fundam.* 15, 230
17. Garside, J., Davey, R.J., 1980, *Chem. Eng. Commun.* 4, 393
18. Daudey, P.J., 1987, *Crystallization of Ammonium Sulphate—Secondary Nucleation and Growth Kinetics in Suspension*, PhD Thesis, Technical University (Delft, Netherlands)
19. Allen, T., 1968, *Particle Size Measurement* (Chapman and Hall, London, UK)
20. Nyvlt, J., 1981, *Chem. prümysl* 31, 172
21. Matz, G., 1982, *Chemie-Technik* 11, 1157
22. Nyvlt, J., Matuchova, M., 1976, *Kristall und Technik* 11, 245
23. Canning, T.F., Randolph, A.D., 1967, Some Aspects of Crystallization Theory: Systems that Violate McCabe's Delta L Law, *AIChE Journal* 13, 1, 5–10
24. Jager, J., de Wolf, S., Klapwijk, W., de Jong, E.J., 1989, A New Design for On-Line CSD Measurements in Product Slurries, *Proceedings of the 10th Symposium on Industrial Crystallization 1987, (ed. J. Nyvlt and S. Zacek)*, Elsevier
25. Brown, D.J., Alexander, K., 1990, On-Line Measurement of Crystal Using a Sample Dilution Method, *Proceedings of the 11th Symposium on Industrial Crystallization 1990*, (ed. A. Mersmann)
26. Söhnel, O., Garside, J., 1992, *Precipitation: basic principles and industrial applications*, Butterworth-Heinemann

27. Davey, R.J., Garside, J., 2000, *From Molecules to Crystallizers: an introduction to crystallization*, Oxford University Press, UK

28. Mersmann, A., Kind, M., 1988, Chemical Engineering Aspects of Precipitation from Solution, *Chem. Eng. Tech.* 11, 264–276

29. Janssen-van-Rosmalen, R., van der Linden, W.H., Dobbinga, E., Visser, D., 1978, The Influence of the Hydrodynamic Environment on the Growth and Formation of Liquid Inclusions in Large Potassium Dihydrogen Phosphate (KDP) Crystals, *Kristall und Technik* 13, 17–23

30. Barkardjiev, I., 1980, Untersuchungen zu Wachstum und Zlichtung von Kristallen aus Kaliumsulfat, *Z. Phys. Chem., Neue Folge*, 122, 187–198

31. Walker, A.C., Kohmann, G.T., 1948, Growing Crystals of Ethylene Diamin Tartrate, *AIEE Trans.* 67, 565

32. Smythe, B.M., 1967, Sucrose Crystal Growth. I. Rate of Crystal Growth in Pure Solutions, *Aust. J. Chem.* 20, 1087–1095

33. Bennema, P., 1966, Technique for Measuring the Rate of Growth of Crystals from Solution in Dependence of the Degree of Supersaturation at Low Supersaturation, *Phys. Stat. Sol.* 17, 555–562

34. Alexandru, H.V., 1971, The Kinetics of Growth and Dissolution of Ammonium Dihydrogen Phosphate Crystals in Solution, *J. Crystal Growth* 10, 151–157

35. Mullin, J.W., Amatavivadhana, A., 1967, Growth Kinetics of Ammonium- and Potassium Dihydrogen Phosphate Crystals, *J. Appl. Chem.* 17, 151–156

36. Mullin, J.W., Garside, J., 1967, The crystallization of Aluminium Potassium Sulphate: Study in the Assessment of Crystallizer Design Data. Part I: Single Crystal Growth Rates, *Trans. Instn. Chem. Engrs.* 45, T285–T290

37. Botsaris, G.D., Denk Jr., E.G., 1970, Growth Rates of Aluminium Potassium Sulphate Crystals in Aqueous Solutions, *Ind. Eng. Chem. Fundam.* 9, 276–283

38. Clontz, N.A., Johnson, R.T., McCabe, W.L., Rousseau, R.W., 1972, Growth of Magnesium Sulphate Heptahydrate Crystals from Solution, *Ind. Eng. Chem. Fundam.* 11, 368–373

39. Phillips, V.R., Mullin, J.W., 1975, A Technique for Observing Growth Layers of Adjacent Faces of a Crystal Growing in Solution, *J. Cryst. Growth* 29, 382–384

40. Tengler, T., Mersmann, A., 1984, Influence of Temperature, Supersaturation and Flow Velocity of Solution on Crystal Growth from Solutions, *Ger. Chem. Eng.* 7, 248–259

41. Kind, M., Mersmann, A., 1990, On the Supersaturation during Mass Crystallization from Solution, *Chem. Eng. Technol.*

42. Schlichting, H., 1982, *Grenzschicht-Theorie*, G. Braun Verlag (Karlsruhe, Germany)

43. Garside, J., Larson, M.A., 1978, Direct Observation of Secondary Nuclei Production, *J. Crystal Growth* 43, 693–704

44. Wang, S., Mersmann A., 1989, *Proc. of the 9th Intern. Conf. on Crystal Growth (ICCG9)*, Sendai, Japan, Elsevier (Amsterdam, The Netherlands)

45. Garside, J., Rusli, I.T., Larson, M.A., 1979, Origin and Size Distribution of Secondary Nuclei, *AIChE J.* 25, 57

46. Berglund, K.A., Kaufmann, E.L., Larson, M.A., 1983, Growth of Contact Nuclei of Potassium Nitrate, *AIChE J.* 29, 867–869

47. Ramanarayaran, K.A., Berglund, K.A., Larson, M.A., 1985, Growth Kinetics in the Presence of Growth Rate Dispersion from Batch Crystallizers, *Chem. Eng. Sci.* 40, 1604–1608

48. Tsukamoto, K., 1983, In-situ Oberservation of Mono-Molecular Growth Steps on Crystals Growing in Aqueous Solution. I., *J. Crystal Growth* 61, 199–209

49. Tsukamoto, K., Sunagawa, I., 1985, In-situ Observation of Mono-Molecular Growth Steps on Crystals Growing in Aqueous Solution: II. Specially Designed Objective Lens and Normanski Prism for In-situ Oberservation by Reflected Light, *J. Crystal Growth* 71, 183–190

50. Kämmer, S., Beckmann, W., Rolfs, J., 1990, Analysis of Surface Features of Potassium Nitrate Using Scanning Tunnelling Microscopy, *Proceedings of the 11th Symposium on Industrial Crystallization 1990, (ed. A. Mersmann)*

51. Seifert, D., 1979, *Untersuchung der Kristallisation von Kaliumchlorid aus wässriger Lösung*, VDI Forschungsheft 591, VDI Verlag (Düsseldorf, Germany)

52. Tengler, T., 1990, *Wachstum und Keimbildung von Ammoniumsulfat bei der Kühlungskristallisation*, PhD Thesis, Technische Universität München (Munich, Germany)

53. Riddiford, A.C., 1966, *Advances in Electrochemistry and Electrochemical Engineering*, Vol. 4, 47, John Wiley & Sons, (eds) P. Delahay and C.W. Tobias

54. Janssen-van-Rosmalen, R., van Leeuwen, C., Smith, J.M., 1970, *J. Crystal Growth* 34, 221

55. Bourne, J.R., Davey, R.J., Gros, H., Hungerbuhler, K., 1976, *J. Crystal Growth* 34, 221

56. Webster, G., Garside, J., 1989, *Industrial Crystallization 87 (ed. J. Nyvlt, S. Zacek)*, 191–194

57. Chianese, A., 1988, *J. Crystal Growth* 91, 39

58. Cochran, W.G., 1934, *Proc. Camb. Phil. Soc.* 30, 365

59. Newman, J.S., 1966, *J. Phys. Chem.* 70, 1327

60. Millsaps, K., Pohlhansen, J., 1952, *J. Aeron. Sci.* 19, 120

61. Norrish, R.S., 1967, *Selected Tables of Physical Properties of Sugar Solutions*, Brit. Food Ind. Res. Assoc.

62. English, A.C., Dole, M., 1950, *J. Am. Chem. Soc.* 72, 3261

63. Webster, G., 1988, *Sucrose Crystal Growth Kinetics Using a Rotating Disc*, PhD Thesis, University of Manchester (Manchester, UK)

64. Nyvlt, J., Söhnel, O., Matuchova, M., Broul, M., 1985, *The Kinetics of Industrial Crystallization*, Academia Praha and Elsevier (Amsterdam, Netherlands)

65. Karel M., Nyvlt, J., 1987, Fluidized bed apparatus for crystal growth rate measurements and its testing with copper sulphate, *Chem. Eng. Commun.* 61, 319–326

66. Langer, H., Offermann, H., 1982, On the Solubility of Sodium Chloride in Water, *J. Crystal Growth*, 60, 389–392

67. Langer, H., 1985, *Zum Stofftransport beim Kristallwachstum aus Lösungen*, Dissertation, RWTH Aachen

68. Ang, H.M., Mullin, J.W., 1979, *Trans. Inst. Chem. Engrs* 57, 237 (London, UK)

69. Nyvlt, J., Vaclavu, V., 1962, *Chem. prümysl* 12, 63 (Praha, Czech Republic)

70. Pekarek, V., Hostomsky, J., Skrivanek, J., 1975, *Kristall u Technik* 10(1) 13

71. Nývlt, J., Pekárek, V., 1980, Crystallization studies by thermometric methods III. The effect of thermal history of solutions on nucleation, *Z. Physik. Chem.* 122, 199

72. Konig Ax., König, An., Emons, H., 1985, *Chem. Techn.* 37, 283

73. Nyvlt, J., 1971, *Industrial Crystallization from Solutions*, Butterworths (London, UK)

74. Nyvlt, J., 1986, Batch cooling crystallization, *Handbook of Heat and Mass Transfer (N.P. Cheremissinoff, Ed.)*, Vol. 2, 1377, Gulf Publ. (Houston, USA)

75. Söhnel, O., Mullin, J.W., 1978, *Chem. Eng. Sci.* 33, 1535

76. Gutwald, T., Mersmann, A., 1990, Batch Cooling Crystallization at Constant Supersaturation: Technique and Experimental Results, *Chem. Eng. Technol.*

77. Randolph, A.D., Larson, M.A., 1988, *Theory of Particulate Processes*, Academic Press (New York, USA)

78. Pohlisch, J., 1987, *Einfluß von mechanischer Beanspruchung und Abrieb auf die Korngrößenverteilung in Kühlungskristallisatoren*, PhD Thesis, Technische Universität München (Munich, Germany)

79. Fleischmann, W., 1985, *Untersuchungen zur Verdrängungskristallisation von Natriumsulfat aus wässriger Läsung mit Methanol*, PhD Thesis, Technische Universität München (Munich, Germany)

80. Jancic, S.J., Garside, J., 1975, On the determination of crystallization kinetics from crystal size distribution data, *Chem. Eng. Sci.* 30, 1299–1301

81. Chianese, A., Sangl, R., Mersmann, A., 1996, On the size distribution of fragments generated by crystal collisions, *Chem. Eng. Comm.* 146, 1–12

82. Gahn, C., 1997, *Die Festigkeit von Kristallen und ihr Einfluß auf die Kinetik in Suspensionskristallisatoren*, PhD Thesis, Technische Universität München (Munich, Germany)

83. Hedström, L., 1994, *Secondary nucleation of pentaerythritol and citric acid monohydrate*, PhD Thesis, TH Stockholm (Sweden)

84. Sangl, R., 1991, *Mechanischer Abrieb von Kristallen als Beitrag zur sekundären Keimbildung*, PhD Thesis, Technische Universität München (Munich, Germany)

85. Wang, S., 1992, *Größenabhängige Wachstumsdispersion von Abriebsteilchen und die Relevanz zur effektiven sekundären Keimbildung*, PhD Thesis, Technische Universität München (Munich, Germany)

86. Van der Heijden, A.E.D.M., 1992, *Secondary nucleation and crystallization kinetics*, PhD Thesis, Kathulieke Universiteit Nijmegen (Nijmegen, The Netherlands)

87. Zacher, U., 1995, *Die Kristallwachstumsdispersion in einem kontinuierlichen Suspensionskristallisator*, PhD Thesis, Technische Universität München (Munich, Germany)

88. Geisler, R., Mersmann, A., Voit, H., 1988, Makro- und Mikromischen im Rührkessel, *Chem. Ing. Tech.* 60, No. 12, 947–955

89. Grootscholten, P.A.M., 1982, *Solid-Liquid Contacting in Industrial Crystallizers and its Influence on Product size Distribution*, PhD Thesis, Delft University of Technology (Netherlands)

90. Mersmann, A., Kind, M., Pohlisch, J., 1986, The Influence of Growth and Nucleation Kinetics on the Mean Crystal Size in Suspension Crystallizers, *Proceedings to World Congress III of Chemical Engineering* (Tokyo, Japan)

91. Mersmann, A., 1986, *Stoffübertragung*, Springer Verlag (Berlin, Germany)

92. Ploss, R., 1990, *Modell zur Kontaktkeimbildung durch Rührer- Kristall- Kollisionen in Leitrohrkristallisatoren*, PhD Thesis, Technische Universität München (Munich, Germany)

93. Marchal, Ph., 1989, *Genie de la Cristallisation Application a l'Acide il Adipique*, PhD Thesis, Institut National Polytechnique de Lorraine Ensic, France

94. Bourne, J.R., Zabelka, M., 1980, The Influences of Gradual Classification on Continuous Crystallization, *Chem. Eng. Sci.* Vol. 35, 533–542

95. Abegg, C.F., Stevens, J.D., Larson, M.A., 1968, Crystal Size Distributions in Continuous Crystallizers when Growth Rate is Size Dependent, *AIChE Journal* Vol. 14, No. 1, 118

96. Heyer, C., 2000, Thesis, Technische Universität München (Munich, Germany)

97. Rumpf, H., 1965, Die Einzelkorn zer kleinerung als Grundlage einer technischen Zerkleinerungs wissenschaft, *Chem. Ing. Techn.* 37(3), 187

98. Pohlisch, J., Mersmann, A., 1988, The Influence of Stress and Attrition on Crystal Size Distribution, *Chem. Eng. Technol.* 11, 40–49

99. Randolph, A.D., 1970, How to Approach the Problems of Crystallization, *Chem. Eng.* 4, 80–96

100. Bhat H.L., Sherwood, J.N., Shripathi, T., 1987, The influence of stress, strain and fracture of crystals on the crystal growth process, *Chem. Eng. Sci.* 42, 609–618

101. Van der Heijden, A.E.D.M., 1992, Growth rate dispersion: the role of lattice strain, *J. Crystal Growth*, 118, 14

102. Gahn, C., Mersmann, A., 1997, Theoretical prediction and experimental determination of attrition rates, *Trans. I. ChemE.* 75A, 125–131

103. Kobayashi, A.S., 1993, *Handbook on experimental mechanics, 2nd ed.* VCH (Cambridge, New York, Weinheim)

104. Lawn, B.R., Wilshaw, T.R., 1993, *Fracture of brittle solids, 2nd Edition*, Cambridge University Press

105. Landolt-Börnstein, 1992, Low Frequency Properties of Dielectric Crystals, *Second and Higher Order Elastic Constants, Vol 29a*, Springer Verlag (Berlin, Heidelberg, New York)

106. Hill, R., 1952, The elastic behaviour of a crystalline aggregate, *Proc. Phy. Soc. London*, A65, 349

107. Wiederhorn, S.M., Hockey, B.J., 1983, Effect of material parameters on the erosion resistance of brittle materials, *J. Mat. Sci.* 18, 766–780

108. Cook, R.F., Pharr, G.M., 1990, Direct observation and analysis of indentation cracking in glasses and ceramics, *J. Am. Ceram. Soc.* 73, 4, 787–817

109. Hutchings, I.M., 1993, Mechanisms of wear in powder technology: a review, *Powder Technol.* 76, 3–13

110. Gahn, C., Mersmann, A., 1995, The brittleness of substances crystallized in industrial processes, *Powder Technol.* 85, 1, 71–81

111. Klein, C., Hurlbut, C. S., 1985, *Manual on mineralogy, 20th ed.*, Wiley (New York, USA)

112. Engelhardt, W. von, Haussühl, S., 1965, Festigkeit und Härte von Kristallen, *Fortschr. Miner.* 42, 1, (1965), 5–49

113. Nadgornyi, E., 1988, Dislocation dynamics and mechanical properties of crystals, Vol. 31, *Progress in Materials Science, Ed. J.W. Christian, P. Haasen, T.B. Massalski*, Pergamon Press (Oxford, UK)

114. Chaudri, M.M., Wells, J.K., Stephens, A., 1981, Dynamic hardness, deformation and fracture of simple ionic crystals at very high rates of strain, *Phil. Mag. A*, 43, 643–664

115. Tabor, D., 1986, Indentation hardness and its measurement: some cautionary comments, in *Microindentation Techniques in Materials Science and Engineering, P. J. Blau and B. R. Lawn eds.*, ASTM STP 889, Philadelphia, PA, 129–159

116. Michell, J.H., 1900, Some elementary distributions of stress in three dimensions, *London Math. Soc. Proc.* 32, 23–57

117. Lawn, B.R., Wilshaw, T.R., 1975, Indentation fracture: principles and application, *J. Mat. Sci.* 10, 1049–1081

118. Gahn, C., Mersmann, A., 1999, Brittle fracture in crystallization processes, *Chem. Eng. Sci.*, 54, 1273–1292

119. Orowan, E., 1949, Fracture and strength of solids, *Rep. Progr. Physics* 12, 185–232

120. Weichert, R., Schönert, K., 1978, Heat generation at the tip of a moving crack, *J. Mech. Phys. Solids*, 26, 151–161

121. Van der Hoek, B., Van der Eerden, J.P., Bennema, P., 1982, Thermodynamical stability for the occurrence of hollow cores caused by stress of line and planar defects, *J. of Crystal Growth*, 56, 621–631

122. Eble, A., 2000, Thesis, Technische Universität München (Munich, Germany)

123. Kind, M., Wellinghoff, G., 1994, Vergleich der Lösungskristallisation organischer und anorganischer Systeme, *Chem. Ing. Tech.* 66, 1064

124. v. Smoluchowski, M., 1917, Versuch einer mathematischen Theorie der Koagulationskinetik kolloidaler Lösungen, *Z. Phys. Chem.* 92, 129

125. Schubert, H., 1998, Thesis, Technische Universität München (Munich, Germany)

126. Nývlt, J., 1983, Induction period of nucleation and metastable zone width, *Collect. Czech. Chem. Commun.* 48, 1977

127. Nývlt, J., Karel, M., 1994, Width of the metastable region in the precipitation of lead iodide, *Collect. Czech Chem. Commun.* 59, 1495

128. Nývlt, J., 1980, Treatment of data on metastable zone width leading to nucleation characteristics, *Kristall u. Technik* 15, 777

129. Giulietti, M., Derenzo, S., Nývlt, J., Ishida, K., 1995, Crystallization of copper sulphate, *Crystal Res. Technol.* 30, 177–183

130. Söhnel, O., Nývlt, J., 1975, Evaluation of experimental data on width of metastable region in aqueous solutions, *Collect. Czech. Chem. Commun.* 40, 511

131. Mullin, J.W., Nývlt, J., 1971, Programmed cooling of batch crystallizers, *Chem. Eng. Sci.* 26, 369

132. Procházka. S., Sùra, J., Nývlt, J., 1977, Temperature programmer and its application in crystallization research, *Chem. list* 71, 1086 (in Czech)

133. Nývlt, J., 1982, The metastable zone width of ammonium aluminium sulphate and mechanisms of secondary nucleation, *Collect. Czech. Chem. Commun.* 47, 1184

134. Karpiñski, P.H., Nývlt, J., 1983, Metastable zone width in salting-out crystallization, *Crystal Res. Technol.* 18, 959

135. Kind, M., Mersmann, A., 1983, Methoden zur Berechnung der homogenen Keimbildung aus wässrigen Lösungen, *Chem. Ing. Tech.* 55, 270

136. Volmer, M., Weber, A., 1926, Keimbildung in übersättigten Lösungen, *Z. Phys. Chem.* 119, 277–301

137. Mersmann, A., 1990, Calculation of interfacial tensions, *J. Cryst. Growth* 102, 841

138. Verwey, E.J.W., Overbeek, J.Th.G., 1948, *Theory of the Stability of Lyophobic Colloids*, Elsevier Publishing Company (Amsterdam, The Netherlands)

139. Glück, B., 1988, *Hydrodynamische und Gasdynamische Rohrströmung: Druckverluste*, VEB Verlag für Bauwesen, Berlin, Germany

140. Nielsen, A., 1964, *Kinetics of Precipitation*, Pergamon Press (Oxford, UK)

141. Angerhöfer, M., 1994, Thesis, Technische Universität München (Munich, Germany)

142. Tavare, N., Garside, J., 1986, Simultaneous estimation of crystal nucleation and crystal growth kinetics from batch experiments, *Chem. Eng. Res. Des.* 64, 109

143. Tavare, N., 1991, Batch Crystallizers, *Rev. Chem. Engng.* 7, 212

144. Aoun, M., 1996, Thesis, Institute National Polytechnique de Lorraine (Nancy, France)

145. Enüstün, B.V., Turkevich, J., 1960, Solubility of fine particles of strontium sulfate, *J. Am. Chem. Soc.* 82, 4502

146. Beckmann, W., Rauls, M.,1987, Zur Genauigkeit der Auswertung fehlerbehafteter Messungen der Wachstumsgeschwindigkeit von Kristallen, *Chem. Ing. Tech.* 59, 422–424

147. Bennema, P., Klein-Haneveld, H.B., 1967, Error Analysis of a Technique to Determine the Rate of Crystal Growth from Solution at Low Supersaturations, *J. Crystal Growth* 1, 225–231

148. Jancic, S.J., Grootscholten, P.A.M., 1983, *Industrial Crystallization*, Kluwer Academic Publishers, (Dondrecht, Netherlands)

149. Mersmann, A., Sangl, R., Kind, M., Pohlisch, J., 1986, Attrition and Secondary Nucleation in Crystallizers, *Chem. Eng. Technol.* 11, 80–88

150. Mersmann, A., 1995, *Crystallization Technology Handbook*, Marcel Dekker (New York, USA)

151. Wöhlk, W., et al., 1982, *MeBanordnungen zur Bestimmung von Kristallwach-stumsgeschwindigkeiten, Fortschritt-Berichte der V Dl Zeitschriften*, Reihe 3, Nr. 71

152. Van Enckevort, W.J.P., 1982, Verification of Crystal Growth Models by Detailed Surface Microtopography and X-ray Diffraction Topography, PhD Thesis, Katholieke Universitert Nijmegen (Nijmegen, Netherlands)

Notation

(Only the main symbols are given here. Many others are defined locally in the text)

Roman symbols	Quantity	Unit
A	area, surface area	m^2
a	surface area per unit volume	$m^2 m^{-3}$
a_i	activity of component i	$kmol\,m^{-3}$
a_T	interfacial area per unit volume	$m^2 m^{-3}$
B	dimensionless driving force	–
B	nucleation rate based on volume	$m^{-3} s^{-1}$
B_o	effective nucleation rate based on volume	$m^{-3} s^{-1}$
C	heat capacity	$J\,K^{-1}$
C_i	mass concentration of component i	$kg\,m^{-3}$
c	molar concentration	$kmol\,m^{-3}$
c_i	molar concentration of component i	$kmol\,m^{-3}$
c_p	specific heat capacity	$J\,kg^{-1}\,K^{-1}$
Δc	concentration driving force or supersaturation $(= c - c^*)$	$kmol\,m^{-3}$
C.V.	coefficient of variation	–
D	diffusion coefficient	$m^2 s^{-1}$
D	diameter of apparatus	m
d	diameter of impeller or tube, distance	m
E_A	activation energy	$J\,mol^{-1}$
F	overall shape factor	–
G	linear crystal growth rate $(= dL/dt)$	$m\,s^{-1}$
g	gravitational acceleration $(g = 9.81\,m\,s^{-2})$	$m\,s^{-2}$
g	overall order of the growth process	–
H	enthalpy	J
ΔH_c	enthalpy of crystallization	$J\,kg^{-1}$
H	height	m
h	specific enthalpy	$J\,K^{-1}$
k_d	mass transfer coefficient	$m\,s^{-1}$
k	Boltzmann constant $(k = 1.381 \cdot 10^{-23}\,J\,K^{-1})$	$J\,K^{-1}$
k	reaction rate constant	var.
k_N	nucleation rate constant	var.
$k_G\ k'_G\ k''_G$	growth rate constants	
k_r	growth integration rate constant	var.
L	particle size	m
L_{50}	mass median crystal size	m
$\tilde{M}$	molar mass	$kg\,kmol^{-1}$
M (L)	cumulative oversize mass distribution	–
M_i	molality	$kmol\,kg^{-1}$
M	mass	kg
$\dot{M}$	mass flow	$kg\,s^{-1}$
m	mass per volume suspension	$kg\,m^{-3}_{sus}$
$\dot{m}$	mass flow density	$kg\,m^{-2}\,s^{-1}$
m_T	suspension density, total crystal mass per unit volume of suspension	$kg\,m^{-3}$
m_i	moments of population density distribution	m^{-i}
N	number of particles per unit suspension volume	m^{-3}
N_{tot}	total number of particles	–

Roman symbols	Quantity	Unit
$\dot{N}$	particles rate per unit volume	$m^{-3}\,s^{-1}$
N_A	Avogadro number ($N_A = 6.023 \cdot 10^{23}\ mol^{-1}$)	mol^{-1}
n	amount of substance	mol
n	population density per unit volume	m^{-4}
n_o	nuclei number density per unit volume	m^{-4}
$n_{o,\ eff}$	effective nuclei number density per unit volume	m^{-4}
P	stirrer power	W
Q	quantity of heat	J
q	quantity of heat per unit mass	$J\,kg^{-1}$
$\dot{Q}$	heat flux	$J\,s^{-1}$
$Q_i(L)$	cumulative undersize distribution	–
$q_i(L)$	undersize distribution, fraction	m^{-1}
$\tilde{R}$	ideal gas constant ($8.314\,J\,mol^{-1}\,K^{-1}$)	$J\,mol^{-1}\,K^{-1}$
R_D	mass dissolution rate	$kg\,m^{-2}\,s^{-1}$
R_G	mass deposition rate	$kg\,m^{-2}\,s^{-1}$
r	radius	m
r	order of the integration process	–
S	supersaturation ratio ($S = c/c^*$)	–
s	impeller speed	s^{-1}
T	absolute temperature	K
$\dot{T}$	cooling rate	$K\,s^{-1}$
t	time	s
u	fluid velocity	$m\,s^{-1}$
V	volume	m^3
$\dot{V}$	volumetric flow rate	$m^3\,s^{-1}$
v	velocity	$m\,s^{-1}$
v_{hkl}	face growth rate	$m\,s^{-1}$
$\bar{v}$	mean face growth rate	$m\,s^{-1}$
$W\,(L)$	mass distribution function	$kg\,m^{-4}$
W_i	mass ratio of two substances	$kg\,kg_{solv}^{-1}$
w	velocity	ms^{-1}
w_i	mass fraction of substance i	$kg\,kg_{sol}^{-1}$
X_i	mole ratio of two substances	$mol\,mol_{solv}^{-1}$
x_i	mole fraction of substance i	$mol\,mol_{sol}^{-1}$

Greek symbols	Quantity	Unit
α	volume shape factor	–
β	surface area shape factor	–
γ	activity coefficient	–
γ_{CL}	interfacial tension	$J\,m^{-2}$
δ	film thickness	m
δ_M	diffusion boundary layer thickness	m
δ_H	hydrodynamic boundary layer thickness	m
ε	local specific power input	$W\,kg^{-1}$
$\bar{\varepsilon}$	mean specific power input	$W\,kg^{-1}$
η	dynamic viscosity	Pa s
η_c	effectiveness factor for crystal growth	–
ϑ	Celsius temperature	$^{\circ}C$

$\Delta\vartheta, \Delta T$	subcooling	K
μ_i	chemical potential of substance i	$J\,mol^{-1}$
ν	kinetic viscosity	$m^2\,s^{-1}$
ν_i	stoichiometric coefficient of i	–
ρ	density	$kg\,m^{-3}$
σ	relative supersaturation $(= \Delta c/c^*)$	–
τ	mean residence time	s
φ	volumetric crystal hold-up	$m^3\,m^{-3}$
ϕ_i	volume fraction of substance i	$m^3\,m^{-3}$
ω	angular velocity	s^{-1}
ψ	sphericity or voidage	–

Subscripts

anh	anhydrous
C	crystallizer
c	crystal
cum	cumulative
dom	dominant, mode value
eff	effective
f	feed
hom	homogenous
hyd	hydrate
i	component i of mixture
L	liquid phase, solution
m	molar value, value per mole
max	maximum value
min	minimum value
p	particle, product
sol	solution
solv	solvent
sus	suspension
tot	total
∞	bulk value
α	beginning
ω	end

Superscripts

| o | standard state |
| * | equilibrium value, saturation |

Dimensionless groups

Damköhler number
$$Da = k_r(C_\infty - C^*)^{r-1} \cdot \frac{(1 - W_\infty)}{k_d}$$

Flow number
$$N_v = \frac{\dot{V}}{s \cdot d^3}$$

Power number
$$P_o = \frac{P}{\rho_{sus} \cdot n^3 \cdot d^5}$$

Reynolds number of particle
$$Re_p = \frac{v \cdot L}{\nu_L}$$

197

Reynolds number
of stirrer

$$\mathrm{Re}_{\text{stirrer}} = \frac{s \cdot d^2 \cdot \rho_L}{\eta_L} = \frac{s \cdot d^2}{\nu_L}$$

Reynolds number of
rotating disk

$$\mathrm{Re}_{\text{disk}} = \frac{\omega \cdot r^2}{\nu_L}$$

Schmidt number

$$\mathrm{Sc} = \frac{\nu_L}{D}$$

Sherwood number

$$\mathrm{Sh} = \frac{k_d L}{D}$$

Relative
supersaturation

$$\sigma = \frac{\Delta c}{c*}$$

Index